# VEDIC PHILOSOPHY

# VEDIC PHILOSOPHY

*Editor-in-Chief*

**Swami Chidatman Jee Maharaj**

**ANMOL PUBLICATIONS PVT. LTD.**

**NEW DELHI - 110 002 (INDIA)**

**ANMOL PUBLICATIONS PVT. LTD.**
**H.O.:** 4374/4B, Ansari Road, Daryaganj,
New Delhi-110 002 (India)
Ph.: 23278000, 23261597
**B.O.:** No. 1015, Ist Main Road, BSK IIIrd Stage
IIIrd Phase, IIIrd Block,
Bangalore - 560 085 (India)
*Visit us at*: www.anmolpublications.com



*Vedic Philosophy*


First Published, 2009
ISBN 978-81-261-3617-9



PRINTED IN INDIA

Printed at Mehra Offset Press, Delhi.

# Contents

# Preface

India colourful and vibrant, a land as diverse as its people. A mosaic of faiths, cultures, customs and languages that blend harmoniously to form a composite whole. One of the world's oldest living civilizations—which gave to the world—the concept of zero, the primordial sound Aum..., Yoga and Buddhism.

She is old. She is young. The two diametrically opposite, different yet unique and dynamic faces of India coexist even at this turn of the millennium. With a history of heritage, learning and civilisation dating back to almost 5000 years or perhaps even more, the modern India presents itself as an ideal country. From snow clad mountains to breezy sparkling blue sea side. From arid deserts with shifting sand dunes to lush green country sides. From ancient relics and architectural ruins dating back into antiquity to modern cities, India has it all.

To the north it is bordered by the world's highest mountain chain, where foothill valleys cover the northernmost of the country's maximum states. Further south, plateaus, tropical rain forests and sandy deserts are bordered by palm fringed beaches. Its southernmost tip is the meeting point of three Oceans.

India is a land of staggering contrast with a mingling of the tradition and modernity that is an unique experience to be in India. One will forever remember the stay in this wonderful Homeland-a land truly of mystery and enchantment. Only India can offer such an astonishing variety of contrasts.

She has given birth to numerous spiritual leaders and founders of some of the popular religions of the present day world. India probably has the most religious diversity in any country. It's the birthplace of Hinduism, Buddhism, Sikhism and Jainism and is among the few places to have a resident

Zoroastrian population. India with her veritable treasure trove of culture, mysticism, philosophy, art, music and architecture to name a few. Much of India's classical music is devotional and the North Indian Hindustani and South Indian Carnatic streams are distinct and both have a complex 'raga' framework. The legacy of dance in India is tremendous. The classical dances of India are numerous such as Kathakali of Kerala, Bharatnatyam of Tamil Nadu, Kuchipudi of Andhra Pradesh, Manipuri and Odissi from Orissa are the prominent dance forms in this country that sways to an altogether novel beat. The Yakshagana, nautanki and puppetry are ancient folk forms that live on till date. The earliest specimens of Indian painting are the ones on the walls of the Ajanta Caves dating back to 2nd century BC. The Mughals had a huge impact on Indian art. The influence of Persian art brought placid garden scenes, illustrations from myths, legends and history into Indian art. Word craft, handicrafts, architecture and sculpture all contribute to this rich and varied domain. Indian literature, both in English and in the vernacular, is ever more popular around the world.

The book aims at presenting perspectives of different aspects of Indian culture. It has topics on Indian classical music and dance, painting, ayurveda and yoga which are traditional and indigenous system of physical and mental wellbeing, language, philosophy and excellence achievements by Indian in the different branches of science and social sciences. It also includes our ancient religious and classic literature.

It is hoped that this book will be welcomed by the general readers as also scholars who are interested in the Indian heritage.

—*Editors*

# 1

# Philosophy of Vedas

The *Vedas* are the main scriptural texts of Hinduism, also known as the Sanatana Dharma, and are a large corpus of texts originating in Ancient India. The Vedas, regarded as *Aaruti* ("that which is heard"), form part of an oral tradition in the form of an ancient teacher-disciple tradition.

As per Hindu tradition the Vedas were 'revealed' to the Rishis referred to in the texts, not composed or written by them. Even though many historians have tried to affix dates to the Vedas there is as yet no common consensus as there is for the scriptures of other religions.

The Vedas are arguably the oldest surviving scriptures in the world. The Vedanta and Mimamsa schools of Hindu philosophy assert that the Vedas are apaurusheya ("unauthored"), that is, they have neither human nor divine origin, and are eternal in nature. As per Hindu tradition, the sage Vedavyasa divided the Vedas into Rigveda, Yajurveda, Samaveda and Atharvaveda at the beginning of the Kali Yuga.

***Overview:*** The sacred books of ancient India. The Sanskrit word veda means "knowledge", more particularly "sacred book". In its widest sense the term designates not only the sacred texts, but also the voluminous theological and philosophical literature attached thereto, the Brahmanas, Aranyakas, Upanishads, and Sutras.

But usually the term veda applies only to the four collections (Samhitas) of hymns and prayers composed for different ritualistic purposes: the Rigveda, Samaveda, Yajurveda, and

Atharvaveda. Of these only the first three were—originally regarded as canonical; the fourth attained to this position after a long struggle. Though differences exist in the language of the four Vedas, still there is such agreement on cardinal points as against later Sanskrit that the term Vedic, which is in common use for the oldest form of the language of India, is amply justified.

***The Rigveda:*** The Rigveda ("Veda of verses"; from ric, or before sonants rig, "laudatory stanza") is the oldest and most important of these collections. In its present form it contains 1028 hymns (including eleven supplementary ones in the eighth book), arranged in ten mandalas (cycles), or books, which vary in extent, only the first and tenth being approximately equal. The poems themselves are of different authorship and date from widely different periods.

The actual date of these ancient scriptures is a nebulous topic. Yet, the description of an extremely cold climate leads some to believe that the Vedas are close to 20,000 years old, but there are some modern scholars who think that the number is exaggerated and should be about 5000. No matter what the age, it is the belief by many these texts were and are the oldest in the world. They express philosophies, realities and truths about life. The texts themselves show that the collection is the result of the work of generations of poets, extending over many centuries. Books II to VII inclusive are each the work of a single poet, or rishi (seer), and his descendants; hence they are aptly called "family books". Book III is attributed to the family of Vishvamitra, IV to that of Vamadeva, V to that of Vasishtha. The hymns in books I and X are all composed by different families. The ninth consists exclusively of hymns addressed to Soma, the deified plant, the juice of which was used for the Soma sacrifice. Books II to VII are the oldest, and book X the most recent, in point of origin.

Hymns to the same deity are usually grouped together. Thus, approximately 500 hymns are addressed to two gods alone: Indra, the god of lightning and storms, and Agni, the god of fire. The term god is actually false, the "gods" (devas) being only highly, elevated prophets, angels, or phenomenons. The

Vedas clearly state that there is only one God, the same as Jews, Muslims, Christians belief alike. One Vedic missionary is the Arya Samaj, its churches are located all over the world, especially in India, the USA, and can teach those who search the Vedas.

The element of nature-worship is a marked feature in most of the hymns, which are invocations of different deities. The value of the great collection as presenting the earliest record of the mythology of an Indo-European people is apparent. Several of the gods go back to the time of Indo-Iranian unity, *e.g.* Yama (the Avestan Yima), Soma (haoma), Mitra (the later Persian Mithra). Some of the divinities, especially the higher ones, still exhibit the attributes which enable us to trace their origin to the personification of natural phenomena.

Thus Indra personified thunder, Agni fire, Varuna the sea, Surya the sun, Ushas the dawn, the Maruts the storm, and others were of a somewhat similar character. Indra was the favourite god of the Vedic Aryans; almost one fourth of all the hymns in the Rigveda are addressed to him and they are among the best in the collection. Next to Indra stands Agni. Many hymns are in honour of Soma.

Other gods invoked are the two Ashins, somewhat resembling the Diocsuri of ancient Greece, the terrible Rudra, Parjanya the rain-god, Vayu the wind-god, Surya the sun-god, Pushan the protector of roads and stray kine. Prayers are also addressed to groups of divinities like the Adityas and the Vishve Devas (all the gods). Only a few hymns sing the praise of Vishnu and of Shiva in his earlier form as Rudra, though these two deities became later the chief gods of the Hindu pantheon. Goddesses play a small part, only Ushas, the goddess of dawn, has some twenty hymns in her honour; these poems are of exceptional literary merit.

The number of secular hymns are small, but many of them are of particular interest. They are of various content. In one (book X, 34) a gambler laments his ill luck at dice and deplores the evil passion that holds him in his grasp. In the same book (X, 18) there occurs a funeral hymn, from which important information may be gained concerning the funeral rites of the Vedic age. Evidently cremation was most in vogue, though burial was also resorted to. There are also some riddles and incantations or prayers exactly like those in the Atharvaveda. Historical references are occasionally found in the so-called danastutis (praises of gifts), which in most cases are not independent poems, but laudatory stanzas appended to some ordinary hymn, and in which the poet gives thanks for generosity shown to him by some prince. Some six or seven hymns deal with cosmogonic speculations.

It is significant that some of the hymns, chiefly in book X, are cast in the form of a dialogue. Here we may possibly discern the beginnings of the Sanskrit drama. The poetry of the Rigveda is neither popular nor primitive, as it has been erroneously considered, but is the production of a refined sacerdotal class and the result of a long period of cultural development. It was intended primarily for use in connection with the Soma sacrifice, and to accompany a ritual, which, though not so complicated as at the time of the Brahmanas, was far from simple.

The Rigveda has come down to us in only one recession, that of the Shakala school. Originally there were several schools: The "Mahabhashya" (great commentary), about the second century B.C., knows of twenty-one, while some later writings know of two only. In these schools the transmission of the hymns was most carefully attended to; a most elaborate mnemonic system was devised to guard against any changes in the sacred text, which has thus come down to us practically without variants.

Editions of the Rigveda were published by Max Muller, "Rigveda-Samhita with the Commentary of Sayancharya" (6 vols., London, 1849-74; 2nd ed., 4 vols., 1890-95); "The Hymns of the Rigveda in the Samhita and Pada Texts" (2nd ed., 2 vols., London, 1877); Aufrecht, "Samhita Text", in Roman characters (2nd ed., Bonn, 1877); selections in Lanman's "Sanskrit Reader" (Boston, 1884); Bothlingk, "Sanskrit-Chrestomathie" (3rd ed., Leipzig, 1897); Windisch, "Zwolf Hymnen des Rigveda", with Sayana's commentary (Leipzig, 1883). Translations were made into: English verse by Griffith (2 vols., Benares, 1896-97);

selections in prose by Max Muller in "Sacred Books of the East", XXXII (Oxford, 1891); continued by Oldenburg, ibidem, XLVI (1897); German verse by Grassmann (2 vols., Leipzig, 1876-77); German prose by Ludwig (6 vols., Prägue, 1876-88). On the Rigveda in general see: Kaegi, "The Rigveda", tr. Arrowsmith (Boston, 1886); Odenberg, "Rigveda", books I-Vi in "Göttinger Gesellschaft der Wissenschaften", new series, XI (Berlin, 1909).

***Righteousness (Dharma) in the Rugveda :*** The Vedic sages created the institution of sacrificial fires (yadnya) as the point of union of God and man, on the earth. Later that very institution became the central focus of the spread of Vedic culture (sanskruti). A sacrificial fire is the very core of social life and the formation of social organisations. According to the Vedic sages The Supreme God exists in the form of a sacrificial fire. Performing sacrificial fires was Their sole code of Righteousness. Somyag is a sacrifice which is accorded the central focus in the Rigveda. Som represents vision and fire symbolises light. It is because of the sun and the rain that all living beings survive. Food too is generated from them. This being the attitude of Vedic sages towards sacrificial fires They linked every important action of life to them.'

***The Samaveda:*** The Samaveda ("Veda of chants") consists of 1549 stanzas, taken entirely (except 75) from the Rigveda, chiefly from books VIII and IX. Its purpose was purely practical, to serve as a textbook for the udgatar or priest who attended the Soma sacrifice. The arrangement of the verses is determined solely by their relation to the rites attending this function.

The hymns were to be sung according to certain fixed melodies; hence the name of the collection. Though only two recensions are known, the number of schools for the veda is known to have been very large. The Samaveda was edited: (with German tr.) by Benfey (Leipzig, 1848); by Satyavrata Samashrami in Bibl. Ind. (Calcutta, 1873); Engl. tr. by Griffith (Benares, 1893).

***The Yajurveda:*** The Yajurveda ("Veda of sacrificial prayers") consists also largely of verses borrowed from the Rigveda. Its purpose was also practical, but, unlike the

Samaveda, it was compiled to apply to the entire sacrificial rite, not merely the Soma offering. There are two recensions of this Veda known as the "Black" and "White" Yajurveda. The origin and meaning of these designations are not clear.

The White Yajurveda contains only the verses and sayings necessary for the sacrifice, while explanations exist in a separate work; the Black incorporates explanations and directions in the work itself, often immediately following the verses. Of the black there are again four recensions, all showing the same arrangement, but differing in many other respects, notably in matters of phonology and accent. By the Hindus the Yajurveda was regarded as the most important of all the Vedas for the practice of the sacrificial rites.

The four recensions of the Yajurveda have been separately edited: (1) "Vajasaneyi Samhita" by Weber (London and Berlin, 1852), tr. Griffith (Benares, 1899); (2) "Taittiriya S." by Weber in "Indische Studien", XI, XII (Berlin, 1871-72); (3) "Maitrayani S." by von Schroeder (Leipzig, 1881-86); (4) "Kathaka S." by von Schroeder (Leipzig, 1900-09).

***The Artharvana-Veda:*** The Artharva-Veda ("Veda of the atharvans or fire priests") differs widely from the other Vedas in that it is not essentially religious in character and not connected with the ritual of the Soma sacrifice. It consists chiefly of a variety of spells and incantations, intended to curse as well as to bless. There are charms against enemies, demons, wizards, harmful animals like snakes, against sickness of man or beast, against the oppressors of Brahmans. But there are also charms of a positive character to obtain benefits, to insure love, happy family-life, health and longevity, protection on journeys, even luck in gambling. Superstitions from primitive ages were evidently current among the masses. To some of the spells remarkably close parallels can be adduced from Germanic and Slavic antiquity.

The Atharvana-Veda is preserved in two recensions, which, though differing in content and arrangement, are of equal extent, comprising 730 hymns and about 6000 stanzas, distributed in twenty books. Many of the verses are taken from the Rigveda without change; a considerable part of the sayings is in prose. The books are of different age; the first thirteen are the oldest, the last two are late additions. Book XX, consisting entirely of hymns in praise of Indra, all taken from the Rigveda, was undoubtedly added to give the Atharvana's connection with the sacrificial ceremonial and thus to insure its recognition as a canonical book. But this recognition was attained only after a considerable lapse of time, and after the period of the Rigveda.

In the "Mahabharata" the canonical character of the Atharvana is distinctly recognized, references to the four Vedas being frequent. Though as a whole this collection must have come into existence later than the Rigveda, much of its material is fully as old and perhaps older.

For the history of religion and civilization it is a document of priceless value. The Atharvana-Veda has been edited by Roth and Whitney (berlin, 1856); Engl. tr. in verse by Griffith (2 vols., Benares, 1897); prose by Bloomfield in "Sacred Books of the East", XLII; by Whitney, revised by Lanman (2 vols., Cambridge, Mass, 1905). Consult Bloomfield, "The Atharaveda" in "Grundriss der Indoarischen Philologie", II (Strasburg, 1899).

***Organization:*** The Mantras are collected into anthologies called Samhitas. There are four Samhitas: the *Rk* (poetry), *Saman* (song), *Yajus* (prayer), and *Atharvan* (a kind of priest). They are commonly referred to as the Rigveda, Samaveda, Yajurveda, and Atharvaveda respectively.

Each Samhita is preserved in a number of versions or recensions (*shakhas*), the differences among them being minor, except in the case of the Yajurveda, where two "White" (*shukla*) recensions contain the Mantras only, while four "Black" (*krishna*) recensions interspersed the Brahmana parts among the Mantras.

The Rigveda contains the oldest part of the corpus, and consists of 1028 hymns. The Samaveda is mostly a rearrangement of the Rigveda for musical rendering. The Yajurveda gives sacrificial prayers and the Atharvaveda gives charms, incantations and magical formulae. In addition to these there are some stray secular material, such as legends.

The next category of texts are the Brahmanas. These are ritual texts that describe in detail the sacrifices in which the Mantras were to be used, as well as comment on the meaning of the sacrificial ritual. Each of the Brahmanas is associated with one of the Samhitas.

The Brahmanas may either form separate texts, or in the case of the Black Yajurveda, can be partly integrated into the text of the Samhita. The most important of the Brahmanas is the Shatapatha Brahmana of the White Yajurveda.

The Aranyakas and Upanishads are theological and philosophical works. They are mystic or spiritual interpretations of the Vedas, and are considered their putative end and essence, and thus known as Vedanta ("the end of the Vedas"). They often form part of the Brahmanas (*e.g.* the Brhadaranyaka Upanishad). They are the basis of the Vedanta school of Darsana.

***Position and Compilation:*** Hindu tradition regards the Vedas as shruti: uncreated, eternal and being revealed to sages (rishis). The rishi Krishna Dwaipayana, better known as Veda Vyasa–"Vyasa" meaning "editor" or "compiler" – reputedly distributed this mass of hymns into the four books of the Vedas, each book being supervised by one of his disciples. Paila arranged the hymns of the Rigveda.

Those that were chanted during religious and social ceremonies were compiled by Vaishampayana under the title *Yajus mantra Samhita* (see Yajurveda). Jaimini is said to have collected hymns that were set to music and melody — "Saman" (see *Samaveda*). The fourth collection of hymns and chants known as the *Atharva Samhita* was collated by Sumanta.

Philosophies and sects that developed in the Indian subcontinent have taken differing positions on the Vedas. In Buddhism and Jainism, the authority of the Veda is repudiated, and both evolved into separate religions.

The sects which did not explicitly reject the Vedas remained followers of the Sanatana Dharma, which is known in modern times as Hinduism.

**'Vedic Shakhas and their Geographical Distribution'**

| *Shakha* | *Samhita* | *Brahmana* | *Aranyaka* | *Upanishad* |
|---|---|---|---|---|
| *Shakala RV* | Oral and Printed | Printed (=Aitareya Brahmana) and Oral | Printed (Aitareya Aranyaka) and Oral | Printed (Aitareya Upanishad) and Oral |
| *Bashkala RV* | Manuscript exists with accents and Padapatha. | — | — | — |
| *Shankhayana RV* | Manuscript exists with accents and Padapatha. | Printed (= Shankhayana Brahmana) | Printed (= Shankhyana Aranyaka) | Printed as a part of the Aranyaka |
| *Kaushitaki RV* | X | Printed (=Kaushitaki Brahmana) and Oral | Manuscript exists | Printed (=Kaushitaki Upanishad) |
| *Ashvalayana RV* | Manuscript exists, with accents and Padapatha. | Same as Shakala | Same as Shakala shakha | Same as Shakala shakha |
| *Paingi RV* | X | Manuscript might exist | 'Rahashya Brahmana' lost. | X |
| *Madhyandina YV* | Oral and Printed | Oral and Printed (Madhyandina Shatapatha) with accents. | Oral and Printed (= Shatapatha XIV.1-8) with accents, | Oral and Printed (= Brihadaranyaka Upanishad = Shatapatha XIV. 3-8) with accents. |

*Contd...*

| *Shakha* | *Samhita* | *Brahmana* | *Aranyaka* | *Upanishad* |
|---|---|---|---|---|
| *Kanva YV* | Oral and Printed | Oral and Printed partially (=Kanva Shatapatha Brahmana) with accents. | Oral and Printed (Book XVII of Kanva Shatapatha) with accents. | Oral and Printed (with numerous commentaries, as 'Brihadaranyaka Upanishad') with accents. |
| *Katyayana YV* | Manuscript (?) | Manuscript (only first 6 books, rest lost) | X | X |
| *Taittiriya YV* | Oral and Printed | Oral and Printed: =Taittiriya Br. and Vadhula Br. (part of Vadhula Srautrasutra) | Oral and Printed (=Taittiriya. Aranyaka) with accents | Oral and Printed (=Taittiriya Upanishad, Mahanarayana Upanishad) with accents. |
| *Maitrayani YV* | Oral (partial?) and Printed | Lost | Oral and Printed (=Maitrayani Aranyaka/ Upanishad) with accents. | Oral and Printed (virtually same as Aranyaka) |

*Contd...*

| *Shakha* | *Samhita* | *Brahmana* | *Aranyaka* | *Upanishad* |
|---|---|---|---|---|
| *Kathaka YV* | Printed. Accents survive only on 1/3 on the text. | Fragments printed and in manuscript) | Printed as 'Katha Aranyaka' (almost the entire text from a solitary manuscript) | Printed as Kathaka Upanishad with numerous commentaries. Katha Shiksha Upanishad also published. |
| *Kapishthala YV* | Printed (from a fragmentary manuscript. Accents left out) | Small fragment extant in manuscript | X | X |
| *Charaka* | Manuscript exists (?) | Lost | In Manuscript | X |
| *Kauthuma SV* | Samhita and Samans printed/recorded | Printed (8 Brahmanas in all). Accents lost long back. | None. The Samhita itself has the 'Aranyaka'. | The famous Chhandogya Upanishad with numerous commentaries |
| *Ranayaniya SV* | Manuscripts of Samhita exist. Samans recorded but not printed | Same as Kauthuma with minor differences. | None. The Samhita itself has the 'Aranyaka'. | Same as Kauthuma. |

*Contd...*

| *Shakha* | *Samhita* | *Brahmana* | *Aranyaka* | *Upanishad* |
|---|---|---|---|---|
| *Jaiminiya / Talavakara SV* | Samhita published. Two distinct styles of Saman recitation, partially recorded and published. | Brahmana published (without accents) – Jaiminiya Brahmana, Arsheya Brahmana | Tamil Nadu version of Talavakara Aranyaka (=Jaiminiya Up. Brahmana) published. | Famous Kena Upanishad |
| *Shatyayana SV* | X (traditions might be similar to Jaminiya SV) | Manuscript | X | X |
| *Paippalada AV* | Two versions: Kashmirian (published) and Orissan (partly published, in manuscript, unaccented) | Lost | X | Famous Prashna Upanishad, Sharabha Upanishad etc. – all published. |
| *Shaunaka AV* | Printed and Oral traditions alive | Fragmentary Gopatha Brahmana (extant and published). No accents. | X | Mundaka Upanishad published. |

***Study:*** Elaborate methods for preserving the text (memorizing by heart instead of writing), subsidiary disciplines (Vedanga), exegetical literature, etc., were developed in the Vedic schools. Sayana, from the 14th century, is known for his elaborate commentaries on the Vedic texts. While much evidence suggests that everyone was equally allowed to study the Vedas and many Vedic "authors" were women, the later dharmashastras, from the Sutra age, dictate that women and Shudras were neither required nor allowed to study the Veda. These dharmashastras regard the study of the Vedas a religious duty of the three upper varnas (Brahmins, Kshatriyas and Vaishyas). In modern times, Vedic studies are crucial in the understanding of Indo-European linguistics, as well as ancient Indian history.

Many forms of Hinduism encourage the Vedic mantras to be interpreted as liberally and as philosophically as possible, unlike the texts of the three Abrahamic religions. In fact, over-literal interpretation of the mantras is actually discouraged, and even the three layers of commentaries (Brahmanas, Aranyakas, and Upanishads), which form an integral part of the Aaruti literature, interpret the seemingly polytheistic, ritualistic, and highly complex Samhitas in a philosophical and metaphorical way to explain the "hidden" concepts of God (Ishwara), the Supreme Being (Brahman) and the soul or the self (Atman). Many Hindus believe that the very sound of the Vedic mantras is purifying for the environment and the human mind. Each may contain a thousand layers of meaning.

***Cosmogony:*** The Vedic view of the world and cosmogony sees one true divine principle self-projecting as the divine word, *Vaak*, 'birthing' the cosmos that we know from 'Hiranyagarbha' or Golden Womb, a primordial sun figure that is equivalent to Surya. The varied gods like Vayu, Indra, Rudra (the Destroyer), Agni (Fire, the sacrifical medium) and the goddess Saraswati (the Divine Word, or Vaak) are just some examples of the myriad aspects of the one underlying nature of the universe.

The word Veda is often derived from 5 roots these days:

- *Vid Jnaane:* To know

- *Vid Sattaayaam*: To be, to endure
- *Vid Labhe*: To obtain
- *Vid Vichaarane*: To consider
- *Vid Chetanaakhyaananiveseshu*: To feel, to tell, to dwell

To these roots is added the suffix 'ghaw' according to Ashtadhyayi 3.3.19, the celebrated text of Sanskrit grammar of Panini. Accordingly, the word Veda means 'the means by which, or in which all persons know, acquire mastery in, deliberate over the various lores or live or subsist upon them.'

Traditionally, the Vedic literature as such signifies a vast body of sacred and esoteric knowledge concerning eternal spiritual truths revealed to sages (Rishis) during intense meditation. They have been accorded the position of revealed scriptures and are revered in Hindu religious tradition. Over the millennia the Vedas have been handed over generation to generation by oral tradition and hence the name "shruti" or "that which is heard". According to tradition they are unauthorised (apaurusheya) and eternal.

The Vedic literature is also called by several other names:

- *Nigama*: Traditional wisdom transmitted from generation to generation
- *Amnaaya*: The root texts or primordial texts of (Hindu) tradition
- *Trayi*: The Vedic texts comprising of Versified mantras, prose mantras, and melodies.

***Dating:*** The Hindus fundamental belief is that the Vedas are sanatan-eternal-and apaurusheya-not composed by human entity. At the beginning of every cosmic cycle of Brahma, Paramatma utters the divine words. Later, at various periods great rishis perceive these divine words and imparted this knowledge orally through generations. It is believed that this Vedic knowledge totally disappeared at the end of Brahmas cosmic cycle and reappeared again in Brahma's next cycle of creation. Later a part of vedic knowledge was written. Maha Rishi Veda Vyas simplified this one Veda by dividing it into four: Rig, Sam, Yajur, Atharva. In Bharatvarshas true tradition the vedas are eternal. In the light of these beliefs, the historical dating of these texts is a concept foreign to Hinduism and the province of secular (academic) philology.

Many historians regard the samhitas as one of the oldest oral traditions. The newest parts of the Vedas are estimated to date to around 500 BC; the oldest text (Rugveda) may be assumed to have been completed orally by 1500 BC after being composed over hundreds of years, but most Indologists agree that a long oral tradition possibly existed before it was written down by the second century B.C. Some parts of the Rigveda as well as the other Vedas are written in Sanskrit.

Some writers have used astronomical references in the Rigveda to date it to as early as the 4th millennium BC. This astronomical evidence has often been claimed to support the out of India theory in relation to origins of the Indo-European languages.

***Etymology:*** The word "veda" means "knowledge", and is derived from the root "vid", Sanskrit for "know", reconstructed as being derived from the Proto-Indo-European root "weid-", meaning "see" or "know". "Weid-" is also the source of the English word "wit", as well as "vision" through Latin. The Czech and Slovak words for "science" are "vìda" resp. "veda", derived from western Slavic "vìdet" resp. "vẹdie" for "know".

***Upavedas:*** The Upavedas are derived from the Vedas and are specific applications of the teachings of the Vedas. The main Upavedas are:

1. *Ayurveda:* Indian healing system, it lays more stress on living with nature instead of fighting it, hence preventive instead of corrective medicine.
2. *Dhanur Veda:* Martial arts.

   Ayurveda and Dhanurveda have points in common. They both work with Marma, or natural Pran (Life Energy) that flows in the body. Ayurveda heals the body, while Dhanurveda is used for defending the body. This concept is also known to Chinese as Acupuncture and related Chinese Martial Arts.

3. *Stahapatya Veda:* Architecture, sculpture and geomancy. Used especially for Temple design.
4. *Gandharv Veda:* Music, poetry and dance.

Some other fields like Jyotish (Indian Astrology), Tantra (based on the Puranas, which are in turn based on Vedas), Shiksha and Vyakara (Grammar and pronunciation) are also based on the Vedas.

***The Six Schools of Vedic Philosophy***

1. *Nyaya:* The Logical School, founded by Gautama
2. *Vaishesika:* Atomic school, founded by Kannada
3. *Samkhya:* Cosmic Principle School, founded by Kapila
4. *Yoga:* Yoga school (includes Raj, Hatha, Kundalini, and Tantra Yoga), supposedly founded by Hiranyagarbha, although the Bhagavad Gita, Patanjali's Yoga Sutras, and Hatha Yoga Pradipika are the most popular treatises that have survived.
5. *Mimamsa (also Purva Mimamsa):* Ritualistic School, founded by Jamini
6. *Vedant (also Uttara Mimasa):* Theological School, founded by Badarayana.

Vedanta was made popular by Adi Shankara also called Shankaracharya, who founded the Swami Order of Monks, and established 4 schools (or Maths) in 4 parts of India to carry on teachings of the Vedas in the 7th century.

## Vedic Thought

***The Wonder of Reincarnation:*** As a river nears the ocean it looks back at its life. The virgin snows on mountain-tops that gave birth to it. The lake down below which was its nursery. The travel through mountain passes where it met its tributaries and gained adulthood. The solitude and tranquility of the forests, the singing of the birds, the lush green valleys laden with wheat, corn and rice. The adventurous ride through cities, gladly accepting their refuse and sometimes flooding them as if in a fit of anger. Thousands of experiences and hundreds of memories. And now its imminent merger into a bigger entity. Its losing of the shores which defined it and gave it its uniqueness.

Merging into the ocean however is just a brief stopover. The water evaporates leaving behind all its impurities in the ocean, the clouds drop snow on a different mountain-top, the snow melts and water feeds into a different river and keeps the never ending cycle going.

Might nature have fashioned the human life and for that matter every type of life in the same manner? A life force appearing in its mortal form, going through its journey, disappearing into a bigger entity and after being cleansed and rejuvenated, reappearing at a different time and at a different place in a new mortal garb.

Does that somehow show us that we should revere the life force within and not indulge in worshipping the external attributes provided to it by Providence? Should this shared life force provide for a common bond between all living beings or should we let ourselves be consumed by our petty differences? And if we choose the latter, would that not be an affront to the very essence of nature?

***Self Observation-I:*** The discovery of self is a must for experiencing depth. It comes by self-observation. The Upanishads describe a simple technique for accomplishing this. They say : deliberately divide your attention at all times so as to direct a portion of it back on yourself. Divide you, the 'person' into an 'observing I' and an 'acting or thinking I'. Within the vast array of selves of your personality, establish an awareness that only watches all the rest. By observing yourself you will realize that not YOU but IT speaks within you, moves, feels, laughs, and cries in you. This concept is enumerated in the following Upanishadic sutra:

Two birds, inseparable companions, perch on the same tree.

*One eats the fruit, the other looks on.*
*The first bird is our individual self,*
*feeding on the pains and pleasures of this world;*
*The other is the universal Self, silently witnessing all.*

*The individual self, immersed in the world of change, deluded, laments its lack of freedom.*
*But when it discovers God, full of dignity and power, it is freed from all its suffering.*

Our Rishis admonished; 'Be present at every breath. Do not let your attention wander for the duration of a single breath.'

Thiourea, a transcendentalist, had this to say about the concept of self-observation: 'I am conscious of the presence of a part of me, which, (as it were,) is not a part of me, but a spectator, sharing no experience, but taking note of it..... When the play, (it may be the tragedy,) of life is over, the spectator goes his way. It was a kind of fiction, a work of the imagination only, so far as he was concerned.

The hallmark of Socrates philosophy was 'Know thyself'.

Discover and nurture the 'observing I', and you will find yourself to be more in command of your life!

***Self Observation-II:*** The art of self observation is practiced by deliberately dividing your attention at all times so as to direct a portion of it back on yourself. Establishing an awareness which does nothing but observe the different selves of your personality, is by no means an easy task. In the beginning, a person forgets to keep the 'observing I' separate from the 'acting or thinking I'. The 'self observing' entity keeps merging into whichever 'I' has control over the person at a given moment. But with persistence, different identities that a person assumes become apparent to him. Through observing them, these selves lose their hold on the person. The 'self-observing' part of 'I' becomes dominant and the person detaches himself from his other identities. This process culminates in his reaching a 'partially awakened' state.

In this 'partially awakened' state, a person does everything with a feeling of awareness and with a certain degree of control. His actions are not mere automatic responses to different stimuli. A person can see himself with complete objectivity. This results in his being able to make more meaningful choices. With continued practice of 'self observation' a person ultimately reaches the highest state-the 'fully awakened' state. In this state a person sees, not only himself but everything around him with full objectivity. The base and meaningless concepts of 'I', 'mine', 'yours' etc. disappear. The constant clamor of the mind is replaced by a sense of serenity. An inner peace prevails and the person becomes 'liberated'.

George I. Gurdjieff, a noted Russian philosopher, who was also a student of Hinduism, found the practice of 'self-observation' to be the starting point of all conscious raising efforts. The concept of 'self-observation' and how to practice it, is the single most revealing gift of the Upanishads.

***Self Observation-III:*** A lot has been written about higher consciousness. It simply means being able to see the things around you, the way they really are, and not have your perception be colored by your personal biases and prejudices. The Upanishadic way to accomplish this is by separating the 'observing I' from the 'acting or thinking I'. A good way to start practicing this technique is to just observe for a few days, your one hand do all it does during your waking hours. Then you can stretch this 'self-observation' to include different parts of your body, ultimately to include your mind. It is a long drawn out process but worth the results it produces.

There are many practical benefits that come from moving your consciousness to a higher level. Extracting and enhancing the 'observing you' from the 'rest of your personality' enables you to 'merely observe your anxieties' rather than be 'part of them'. That is a great way to reduce stress. Self-observation reveals to you how small a portion of your most valuable possession your emotional energy is expended wisely. People normally squander it on a host of utterly futile activities such as anxiety, purely imaginary fears, agitation and anger, and also on maintenance of an exhausting degree of muscle tension. Having discovered how wasteful you are in the use of your nervous energy, you can improve your daily life by using it wisely.

By moving to a higher level of consciousness, you can see things from other people's viewpoint. This gives you an edge in resolving minor disputes before they flare up into full-scale egotistical wars. By conserving your reservoir of emotional energies, by separating your 'self' from 'the crises in your life' and by having your perceptions devoid of any biases, your life becomes a lot simpler. The choices you make in life are more thought out. The mental equilibrium you attain calms you down. It has been said that 'awakening is the evolutionary destiny of mankind'. Why not embark on this journey now?

***Man—The Ultimate God:*** Sadhna connotes the successful achieving of a desired end. It is the instrument for the attainment of Siddhi (perfection). The first step in sadhna is the desire for liberation (mumukshutva). Man alone possesses the desire to attain perfection. The gods are already divine and the animal kingdom does not have the desire for perfection. Only man has the conscious urge to expand without limits, to reach out and touch the stars.

Based on his study of the Hindu scriptures done in India, Troy Organ, a professor of Philosophy at Ohio University, wrote, "Man is a real living, growing entity while god is an ideal being. Man and god are identical in essence, but different in form. Man is real potentiality; god is ideal fulfillment. Man is to be fulfilled as god is fulfilled, but not like god's fulfillment, for god's fulfillment is a static fulfillment — there are no possibilities in gods. Man is to be perfected beyond god, for he is to be perfected in dynamic reality. The attainment of god-realization is not the ideal goal of human life. The Upanishads claim that the Self, the Atman, not god is to be realized. Various gods demythologized are merely the symbols of the full realization of human potentialities."

Self observation leads to awakening. An Awakened man, a man who has pushed himself to the limits of possibilities, one who has achieved perfection in dynamic reality under the constraints of time and space, represents the highest value. That is why the Puranas say that the life of man is desired even by gods of heaven, since it is only through a human incarnation that final liberation can be achieved.

Our purpose on this planet is not just to populate the place and in the words of Gibran, 'to fulfil life's longing for itself', but to awaken and actualize the potentialities present only in man.

And when a man attains Siddhi through Sadhna in dynamic reality, he becomes one with Brahman. He becomes the ultimate god.

***Human Neurosis:*** Our desire to understand and rationalize the infinite with a primitive intellect is the most amazing thing about us humans. Science enables us to analyze and validate events happening in the material realm; it lets us observe activities and processes in a controlled environment in a lab, but to extend its application to the spiritual plane is sheer madness. How could one measure the size of something infinite with a finite length of tape? How could eye see the invisible spirit? How could an analytical intellect mired in reasoning, comprehend the divine? It is simply ego driven neurosis that makes a spiritually unhinged person to expect to observe, study and validate spiritual phenomenon with his limited intellect.

***Bhakti:*** When the intellectual curiosity has been satiated, when there are no more questions to be asked, when all the spiritual knowledge catalogued and presented in flowery prose by the learned people has been scanned, dissected and assimilated; without the person being able to still the tidal waves of his psyche, it is then that the mental speculation gives way to the practice of 'Bhakti'. It is then that the intellect surrenders to the divine in a spirit of devotion. It is then that the man's ascent to a spiritual life begins. A person introduces an element of 'Bhakti' in his life with his very first visit to a temple. When he gets there, the person does not have to recite any prayers, he does not even have to utter a single word, for the Lord can see the devotion in his heart. Just showing up with folded hands and devotion in his heart for a *Darshan*-Darshan, which requires neither prayer nor commitment but rather is an act of being in His presence, an act that submerges the self in a striving for the infinite-stills the troubled waves of ones mind. About a visit to the temple, the poet says,

*"kahne kee zaroorat nahin, aana hee bahut hai,*

*is dar pe tera sis jhukana hee bahut hai"*

The practice of spirituality through devotion provides the necessary filter which takes the sting out of the harsh realities of human existence and makes life bearable.

***Rhythms of Life:*** Human nature dictates that we find an anchor to build our lives around. In our desperate search we go from door to door, we bounce from person to person and we jump from coattail to coattail to find somethin stable, someone we could depend on, in times of crisis. At times we feel that we have found that person and we latch on to him. When bad times do come around, as they must, we are in a state of disbelief when we find ourselves bereft of the one person we focused our lives on, one person we centered our pride on, one person who was supposed to be our anchor. Suddenly the framework of our existence seems to be receding from us. We are left only with shattered dreams.

A person who lacks a stable anchor cannot maintain equilibrium in the face of adversity. And as we well know, human relations are all transitory. We have to find something more stable, someone who would not disappoint us in times of need. That is why we have to cultivate a consciousness in which the omnipotent Lord is the anchor. The fortieth chapter of the Yajur Ved states that we should build our lives around Him, as He is the only one who is eternal and would stand by us in every adversity. To start with, one could simply visualize the deity of Lord Krishan and chant the Hare Krishan Mahamantar any time he is not mentally occupied with doing something. Over time this could extend to keeping that image of the Lord in the subconscious at all times. When He rules the subconscious mind, the person is surrendered and the rhythms of life are keyed to performing His service. In the Bhagwad Gita (12.8), Lord Krishan says, "Just fix your mind upon Me, and engage all your intelligence in me. Thus you will live in Me always, without a doubt".

Continued practice in surrendering himself enables a person to develop the consciousness capable of remembering the Lord at the time of final exit. In the Bhagwad Gita (8.5), Lord Krishan says, "Whoever at the end of his life quits his body remembering Me, attains My nature. Of this there is no doubt". Mahatma Gandhi did not suddenly think of invoking the Lord's name by saying *Hey Ram* at the time of his death. He spent his whole life with Him as his anchor. He constantly practiced keeping Him in his subconscious. And so must we....so that our faith becomes a living experience.

***Arjun's Paralysis:*** Many scholars have studied the Bhagwad Gita and tried to provide us with an insight into the teaching of Lord Krishan. Although every chapter of Gita alludes to some aspect of our moral code, the most powerful teaching arises from Arjun's paralysis on the battlefield. Arjun's perplexity arises over the question of dharam, which as a warrior, bids him to fight in order to protect the goodness, and as a person, forbids him to kill the members of his own family, his revered teachers and dear friends. Faced with Arjun's renunciation and his inability to wage war, Lord Krishan resolves his dilemma in two parts. He reminds Arjun that: (1) he must observe his dharam and, (2) the soul survives the body. In the first instance He tells Arjun that he must perform his duty, which for a warrior is to fight the evil forces to the very end and preserve goodness. Secondly, one should not worry about death, as the soul can never be slain. Rather it sheds one body at death and takes on a new one, in birth after birth.

Further Lord Krishan says that successive rebirths may be avoided by observing your dharam with complete disregard to the fruits of your action. He exhorts Arjun to hold alike happiness and unhappiness, gain and loss, victory and defeat in pursuit of his dharam. A person need not become an ascetic to achieve liberation, but only renounce the fruits of his action, and perform his duty without caring about the results. Lord Krishan says that your entitlement is only to the action, and not to the fruit it bears.

These are the answers to Arjun's dilemma. But there is another teaching which is of the greatest importance. This is the teaching of devotion to God. Lord Krishan says that he who along with his struggle for survival, also remembers Him and adores Him, is the greatest ascetic. Steeped in this Hindu philosophy, one overcomes all obstacles in life and leads a peaceful and happy life.

***The Yoga of Devotion:*** In the twelfth chapter of the Bhagwad Gita, Arjun enquires from Lord Krishan about the relative merits of reaching Him through devotional service versus through worshipping the impersonal Brahman, the unmanifested.

Lord Krishan answers, "Those whose minds are fixed on me in steadfast love, worshipping me with absolute faith, I consider them to be most perfect." He further adds, "As for others who worship the unmanifest, indefinable and changeless, that which is omnipresent, constant, eternal and lies beyond the perception of the senses, and they do so by holding all the senses in check, are tranquil minded and devoted to the welfare of humanity, and see the Atman in every creature; they will also come to me."

"But the devotees of the unmanifest have a harder task, because the unmanifest is very difficult for embodied souls to realize."

Bhakti yoga, or the process of reaching Him by being in direct devotional service to the Lord is easier and natural for us humans. The individual soul is embodied since time immemorial, and it is very difficult for a common man to visualize that he is not the body and that the Lord does not have any physical attributes either. In Bhakti yoga, as described in Bhagwad Gita, a person accepts the Deity of Krishan as Brahman with body features and engages himself in devotional service to the Lord. So it becomes a very easy and natural process for him to reach the Supreme Being.

Janana yoga, or the process of reaching Him through understanding the unmanifested Supreme Being is very difficult. By his very nature, a common man has trouble identifying with the formless Supreme. However, if he is persistent in his quest, he might realize the eternal and attain spiritual realization through the guidance of a highly learned devotee.

But the unguided intellectualization of the unmanifest may lead a person to become an atheist. Unfortunately a large percentage of today's intelligensia falls squarely into this category.

Bhakti Yoga, according to the Bhagwad Gita, is the shortest path to spiritual realization.

***Modes of Material Nature:*** The last few chapters of the Bhagwad Gita enumerate the modes of material nature and the characteristics of people living under their influence.

The three modes of material nature are 'sattvaguna' or the mode of goodness, 'rajasguna' or the mode of passion, and 'tamasguna' or the mode of ignorance. Lord Krishan says, "Sattva rules a person who offers sacrifices in accordance with scriptural instructions and does not covet their fruits. Sattoguni, or a person inspired by sattva is impelled by an inner sense of duty. The performance of sacrifices by a Rajoguni is for the outward show, in the hope of a divine reward and is inspired by rajas. The Tamogunis on the other hand, totally disregard the scriptural instructions and make no offering, no prayer of dedication, no gift to the priest and are devoid of faith."

The Lord further adds, "The austerity of the body comprises of reverence for the gods, brahmins and gurus; uprightness, physical cleanliness and sexual purity and nonviolence. Austerity of speech comprises of speech that does not hurt anyone, is truthful, kind and beneficial, as well as daily recitation of scriptures. The austerity of mind comprises serenity of mind, gentleness, self control and inner purity. The triple austerities practiced with the highest faith and with nary a thought of reward crossing ones mind is the nature of sattva. The austerity undertaken out of self-pride, and in order to gain the reputation and homage attendant on pious acts is of the rajas type. Austerity practiced under some foolish misconceptions, by means of self torture or to hurt another person is of the tamas type."

Clearly a satto-guni will do everything as an offering to God. Any work done in this mode shall be the result of doer's best effort and will be performed with great willpower and determination. The motive for the effort will be something higher than the next paycheck or the next promotion. Internally a person will be all fired up to perform but outwardly he would appear to be calm and unstressed as there is no passion involved in the effort. The non-attachment to rewards of ones actions is sometimes considered to suggest coldness and lack of enthusiasm. But in reality, one is freed from the fear of failure and the desire of rewards, and offers everything he does as a sacrament of devotion to his duty. Leading ones life as a satto-guni therefore, is the most spiritual and stressfree mode that one should strive to achieve.

***Householder:*** In the Bhagwad Gita, Lord Krishan says that he who along with his struggle for survival remembers Him and chants His name is the biggest ascetic. He clearly puts a spiritually devoted householder ahead of any other group of people. This is very well illustrated by the following incident.

It is said that once Sage Narad and a common householder entered the court of Lord Krishan almost at the same time. Lord Krishan chose to see the householder first. Now Sage Narad considered himself to be the foremost devotee of the Lord. So being passed up in favor of a common householder really hurt him. So much so that when he had an audience with the Lord, he complained about this incident. The Lord decided to teach the Sage a lesson. He handed him a pot full of 'ghee' and asked him to put it on his head and to go around the world without spilling a drop of it. This, the Sage was able to accomplish without any problem.

So when he came back to the Lord after going around the world, the Lord commended him on his feat and asked him as to how many times did he chant His name during the trip. The Sage said that the Lord must be really joking for he was so busy balancing the pot of ghee that he could not afford to be distracted even for a single instant. Lord Krishan told him that the house holder had to juggle a thousand responsibilities that go with being a father, a husband, a son, a brother, a breadwinner and still found time to chant His name. Sage Narad on the other hand, had only the task of balancing a pot of ghee, and could not find time to chant His name. Therefore the householder was leading a more spiritual life than the Sage. The Sage felt humbled and recognized the greatness of a householder.

Householders are the backbone of the human race. They are the ones who grow the food we eat, design and manufacture consumer goods and appliances needed to make life livable, and provide health care to get us back on our feet when we get sick. They are the ones who contribute most to the material development of society since they have a vested interest in seeing that their children inherit a better world than the one in which they were raised. In short a householder is the most important element of the human race and spiritual householders are the beings most loved by God.

*jag to andhakaar hai, is se tu nikal...*

*gyan kaa prakash kar, mod tu samabhal...*

*Radha Krishan ke charan main nitya kar naman...*

*Man to mandir hai......haridaya hai vrindavan....*

***Temple Rituals-A Psychological Analysis:*** A person's quest for spirituality starts simply with his visiting a temple. To facilitate the process of his spiritual growth, some temples require their devotees to adhere to a dress code. The temple walls are decorated with religious symbols and images. The 'puja' itself consists of singing of hymns, bhajans etc. followed by a 'parvachan' or a class on religion/spirituality; with 'Aarti' being the final offering to the gods. There is a psychological reasoning behind all this. The dress code, all the temple decorations, the aroma of incense, the fresh flowers, the nicely dressed deities on the alter; are all meant to create an atmosphere conducive to a spiritual experience. They are the first step of a process meant to alter the state of one's mind, to bring it to a state where the internal turmoil subsides for a duration. The spiritual vibrations of the temple atmosphere lift the soul to a level where it becomes capable of connecting with the divinity. It essentially prepares the person for what he has come there for.

In our everyday existence, we hide our innermost feelings and wear a shield against any intruding thoughts. Every bit of information is thoroughly screened by our rational mind before it is accepted or rejected. But if we are to learn anything from a spiritual master, the sieve has to be temporarily put away. Subliminal thoughts have to penetrate the mind in their entirety, without being filtered or diluted in any way.

The second step in creating the proper state of mind is to remove the filter separating the ears from the brain, to give our defense mechanism a rest. This is done by singing bhajans and hymns. The mind, through this exercise gets into a voluntary receptive mode. It is then, that the brain can be programmed with the noble thoughts-be they to sing the glory of the Lord or ways to deal with our everyday problems in a conscientious manner. A spiritual master can be most successful in the healing process of a mind beset by the badgering of the trivialities of life; he can convey the scriptural philosophy on virtous living, only to a receptive mind. And the mind is made most receptive when its guards are taken down, when it stops fighting the sensory signals coming in, when there are no obstructions between the sensory perceptions and memory cells.

The decoration of the temple with divine symbols and images, the incense, the flowers, enforcing a proper dress code, singing of the hymns and bhajans are all a prelude to the ultimate goal-to submit oneself to the reprogramming of ones brain with the doctrines of noble living.

***Renew Thyself:*** The Vedas say, 'All intelligences awake with the morning.' "Being awake is not just being physically up and around but being men tally vigorous to celebrate another day of living "observed Thiourea, "The millions are awake enough for physical labour; but only one in a million is awake enough for effective intellectual exertion, only one in a hundred million to the poetic or divine life. To be awake is to be alive. I have never yet met a man who was quite awake. How could I have looked him in the face?"

As we grow older, some of us get lost in the monotony of life. We tend to lose our zest for living. As Thiourea observed, "the morning ceases to be the most enjoyable season of the day". We are awakened not by our newly acquired force and aspirations from within but by some electronic gadgets. It is not the undulations of celestial music that wake us up but the prolonged chatter of a radio. And that is very sad; because it signals the decline of our excitement for life.

To quote Henry David again, "A person who is physically, mentally and spiritually awake in the morning and who keeps pace with the sun; for him the whole day is a perpetual morning. It does not matter what the clock says or what the people around him say, morning is when he is awake and there is dawn in him. And if he can keep the spirit of early morning with him throughout the day, then he has lived through, not a day but a long morning". One must be drunk with living and not ever let himself be resigned to life. If one finds oneself at a dead end, he should develop some new avenues for the release of his creative energies. It does not have to be a money making endeavor. One has to develop a state of mind in which he is alert and buoyant throughout the day and the whole day becomes one long morning.

"Renew thyself completely each day; do it again, and again, and forever again". Keep the spirit of Goddess USHA alive in you.

***Be A Light Into Yourself*:** Hindu religion is based not on believing a certain doctrine or dogma, but in realizing the Divine. It is based not merely on being loyal to a faith but on striving to become one with the infinite. And there is a definite progression that a seeker goes through. A beginner might need symbols or pictures to focus his attention, and to stop his mind from wandering while in prayers. But if he is earnest about his spiritual development, he would soon progress to a stage where he would not need any props for performing his sadhna to reach the infinite. In his quest for the eternal, he would advance from material worship to mental worship and then on to the highest stage where he can realize the divine.

One must not be content with accumulating scriptural knowledge. That is just the beginning of a long journey. The discovery of inwardness is the essential basis of spiritual life. The scriptures could point out the road but each man must travel it for himself. Reading about a past buddha's communion with God is not going to result in your communion. It can excite you and inspire you to have the same experience. But in the end you must travel alone. You must find your own salvation. "For thousands who talk, one can think; for thousands who think, perhaps one sees and understands". Use the talking and thinking stages as stepping stones to rise to the ultimate level of experiencing the Divine.

Those of us who lead outward lives without being touched to our inner depths, do not understand life itself. We believe that we do our duty to religion by accepting the letter of faith and making a token cash donation to the temple of our choice. This results in our spiritual dependence and forces us to accept what others say about the religious truth. But once the individual in his freedom of spirit pursues truth and builds up a center in himself, he has enough strength and stability to deal with all that happens to him. He has the ability to fight back and retain his peace even when he is faced with adverse conditions. Liberation comes by experiencing the truth on your own. It comes not from accumulation of information, but from inner transformation.

Gautama Buddha said to his disciples, "Be a light unto yourself." Find the truth, build your own light and let it shine your path. Remember, the goal is not to find out how others attained bliss but to use that knowledge to attain it for yourself.

## Abhidharma

Buddhism has at its core a psychology little known to the adherents but quite familiar to the Buddhist monks. 'Abhidharma' or the 'ultimate doctrine' has as its basis one of the most systematic psychologies of the world and elaborates Gautama Buddha's penetrating insights into human nature. This is the practical psychology that devotees apply to discipline their minds and hearts in order to attain a more ideal state of being.

According to Abhidharma, the human personality is like a river that keeps a constant form, seemingly a single identity, though not a single drop is the same as a moment ago. Each successive moment of our awareness is shaped by the previous moment, and will in turn determine the following moment. Each mental state is composed of a set of values, called mental factors, that combine to define that mental state. One's mental state at any given moment is a function of the biological and situational influences, in addition to a carryover from the preceding psychological moment.

The mental factors could be healthy ones or unhealthy ones. Good mental health depends on our ability to plant healthy factors in our mind so that they can inhibit and suppress unhealthy factors. The central healthy factor is 'Insight' which implies an ability to 'clearly perceive an object as it really is'. 'Mindfulness' is the 'continued clear comprehension' of an object. The two of them together 'perceive' and 'hold' clarity in one's mind and are sufficient to suppress all unhealthy factors.

The formal Western psychology, which is only about a hundred years old is merely a recent version of the human endeavor undertaken centuries ago in India. Modern psychology has rephrased the Buddhistic philosophy by saying that 'we should balance the negatives in our lives by the positives. We hould try to keep everything under proper perspective. We should remember that what is an obstacle for one person is an opportunity for another; what one man calls a stumbling block is used as a stepping stone by another.'

There is a humorous story which illustrates the concept of keeping a positive attitude. Two hunters were offered a bounty of $5,000 for each wolf that they could trap alive. For days they searched the mountains and forests in their area looking for the elusive prey. Exhausted one night, they fell asleep dreaming of getting rich one day. In the morning, one hunter opened his eyes, only to find that they were surrounded by wolves with flaming eyes and bared teeth. He nudged his friend with great excitement and said, "We are finally rich".

In the history of mankind there have been a few famous people and vast multitudes of ordinary folks, and the true winners among them have been the ones who kept a positive attitude toward life. Ralph Waldo Emerson wrote, "Don't waste yourself in rejection, but chant the beauty of the good".

Neutralizing the negative thoughts and replacing them with positive thoughts was good advice some 2500 years ago, and it is good advice today. Crush a negative thought today and replace it with a positive one. Experience the peace and happiness that flows from this process. Go ahead, do it now!

***Heaven—Where Does it Exist?:*** The concept of a heaven in the hereafter is perhaps as old as the human race itself. But in the entire history of mankind if one looks at the definition of heaven of any culture, of any religion, of any race in an era; one finds that it consists of the things that the people in that group are deprived of, or the things that they really miss in their lives. They project these things to be present in abundance in 'their heaven'. As an example, the Indian heaven almost invariably has rivers of milk and 'ghee' flowing through it. It is nothing but a longing for three sumptuous meals a day, which lot of people don't get. When there is shortage, even of clean drinking water, you can well under stand why we would imagine heaven to be loaded with food.

The second thing that almost invariably pops up in the definition of heaven is that it is peaceful and tranquil up there. Might that reflect our desire for an end to the hatred that different ethnic and religious groups harbour against each other?

With this thought in mind, should it not be the aim of life to create the conditions of heaven right here on earth? Should we not be working to achieve material and spiritual prosperity in our lives and thereby make the whole concept of heaven in the hereafter obsolete? If we could tame our carvings and make our lives full of joy, we probably would not even accept the concept of heaven, much less long to retire to. And that brings us to the idea of feeling happy with what we have; striving for more but nevertheless being at peace with ourselves. The Upanishads tell us to think of ourselves as mere users and not owners of our worldly possessions, of our status in the society and of the family we are blessed with. Following this canon just might do the trick and enable us to experience the heaven while we still reside on the mortal shores.

***You Belong in the Marketplace:*** The restless and incessant striving, with joy in the strife, is the greatest happiness that a man can experience in this world. The Isha Upanishad states that one must will himself to live for at least a hundred years and be occupied every minute of it. And the love of strife, the constant meeting of challenges head on, is what gives meaning to life and makes one desirous of living a long life. The poet says, "Zindagi har kadam ek nayee jung hai" (life is a struggle at every step of the way).

There are days when a person gets up and feels like crawling right back into the bed. For some people such days out-number the times when they are excited about another day of living. They hope to strike it rich one day through winning a lottery or through inheritance and then be able to live happily ever after without having to struggle every day to make a living. They somehow give up on life. Again the poet says, "Hai zindagi bhee maut jo himmat ko haren hain" (people who have lost the courage to fight are the living dead) A person who gives up the struggle, also loses his will to live on. For him the aspirations of youth end up in the disillusions of old age.

Whatever be your vocation or your hobby, you belong in the hustle and bustle of the marketplace. You belong where action is, because that is what sustains life. Designing new gadgets and fixing the old ones, treating and nursing sick people, being engaged in spiritual activities, making business deals, acquiring and disseminating knowledge, creating art and music, doing volunteer work, being engrossed in leisure activities is what you should be occupied with. As water retains its clarity only when it flows in a stream and becomes muddy and stale when it loses its flow; So does life lose its meaning when there is no struggle, no battles to fight and no excitement to keep the will to live strong. A person stays tuned up and razor sharp by being in the markeet place.

It is the struggle, and not the goal that brings salvation to the soul. The moment of man's greatest happiness is the moment of striving for others. And to make his strife more meaningful, man must over soar his own little self, and must join with others in some united effort for the greater good of humanity. Above all he must stay involved.

***Death Penalty:*** Every time the state executes a convicted criminal, the debate over capital punishment resurfaces. The collective moral conscience of the society, which decides the fate of these individuals is largely shaped by our religious beliefs. What does our belief system dictate? Consider the following two scenarios from our scriptures:

1. In Mahabharat, at one stage in the great battle of Kurukheshter, Arjun was pitted against Karan. During this fateful fight, one wheel of Karan's chariot got stuck in the bloody mire of the ground. As he descended to

make his chariot mobile again he requested Arjun to hold his arrows while he attended to his chariot. That was a reasonable request as per the norms of war fare in that age. But Shri Krishan urged Arjun to continue the fight. Arjun paused and hesitated since he did not want to take advantage of this awkward situation, but Shri Krishan once again exhorted him to continue fighting and to kill Karan. At this Arjun shot an arrow on his bow Gandiva, which cut off Karan's head.

2. In Ramayan, Lord Shiva is said to have bestowed a special protective power on the evil king Bali. Under the umbrella of this protective power, no one could defeat him in a fair fight. As per the norms of the day, for a fight to be fair the two dueling opponents had to fight face to face. No sneak attacks were permitted. Even though Bali was bitterly detested by his subjects, yet no one could penetrate his protective shield in a fair fight and dispose him off. It was with this dilemma that Bali's brother Sugreev enlisted Shri Ram's help. Upon hearing the facts, Shri Ram did not hesitate in setting aside the rules of fair play and killed Bali with an arrow shot from behind a tree. There are several other instances in our scriptures in which even God Himself, in His different human incarnations bent the rules of fair play in order to get rid of evil. All these instances seem to point in the same direction, that is, 'Fight an evil per son to the end and take advantage of his every weakness to annihilate him. Show no mercy and even suspend the code of honour temporarily, to deal with the situation.' Can this code be extrapolated to mean that our scriptures endorse death penalty? You be the judge.

***To Pray or Not to Pray:*** Secularists point to the rivalries between different religions as a proof of the futility of religion. Scientists question the very existence of God. Common logic dictates that when you pray, there is no one out there listening to you. The words you utter during your prayers just get lost in thin air. Religious people insist on the existence of God. So who is right, and who is wrong? Who can provide an answer to this dilemma about the very existence of God and about the need for prayers in our daily lives?

Psychologists seem to have a valid explanation. They think that self-talk is the way to propel yourself in a certain direction. For example when you repeatedly tell yourself that you need a new car, you subconsciously programme your brain into working in that direction. Once the idea gets firmly implanted, your brain implores you to start looking for ways to finance your purchase. It might take the shape of another part time job. Or it might result in bigger savings from your present income. Whatever the mode your brain directs you to accumulate enough funds for your desired acquisitions.

People have been known to recover from near fatal diseases like cancer by asserting their willingness to live on. By their selftalk they were able to persuade their brain to manufacture and release the drugs required to kill the organisms attacking their bodies. Granted that it does not happen in every case. But people have been known to have willed themselves few extra years after the medical profession had given up on them. In each of these cases, the will to live on, the self-talk to programme their brain into fighting their malady played an important part in their healing process.

Psychologists liken prayer to self-talk. They think that the self-talk programmes our brain which in turn directs our actions towards achieving our desired goals. Priests on the other hand, look at it from a religious point of view and insist that God does listen to our prayers and grants us that; which we are worthy of. Whichever point of view you may agree with the one put forth by the psychologists or the one put forth by the priests the final conclusion is the same. Prayers are a very important part of our daily lives.

In the words of late Dr. Radhakrishnan, "When in the sorrow of death or the suffering of despair, when trust is betrayed or love desecrated, when life becomes tasteless and meaningless, man stretches forth his hands to heaven to know if perchance there is an answering presence behind the dark clouds; it is then that he comes into touch with the supreme in the solitude of his consciousness, in the realm of the profound and intense. It is the world of light and love in which there is no language but that of silence. It is the world of joy that reveals itself in innumerable forms." Must we wait for moments of despair before we acknowledge His presence? Must we be driven to total defeat before we surrender? Must we feel helpless before we start to pray?

***Synchronicity:*** A friend of mine, Mr. S. had gone to Calcutta to look for a bride. While in India, he happened to travel to Bombay. There, in a store Mr. S met Mrs. M who was also visiting from the U.S. and had been looking for a husband for her younger sister Miss P. Mrs. M invited Mr. S to her parents' house. The next day Mr. S and his parents went to meet his prospective bride and her family and the rest, as they say is history. This is a true story which happened in the early seventies.

The pleasant ending of this story was the culmination of a series of coincidences. Mr. S and Mrs. M visiting India at the same time, Mr. S travelling to Bombay, his chance encounter with Mrs. M in a store, leading to his marriage to Miss P. To us Hindus, it was fate, divine intervention or simply a matter of Sanjog. Our philosophic thought teaches us about the acceptance of life as it unfolds because we believe that the Supreme Power has a hand in everything that happens to us.

In the West, the concept of Sanjog, until recently had been an alien idea. But psychoanalyst Carl Jung was fascinated by his study of this phenomenon in which two events take place at the same time with neither one causing the other to hap pen, and yet they are related to one another in a meaningful way. He coined a new name for it-Synchronicity. Thus was another ancient Hindu tenet blessed by the scientific West. It is not looked at any more as just another happening based on statistical probability.

The coming together by apparent chance, of factors that are not causally linked but that nevertheless show themselves to be meaningfully related is at the very heart of the process by which the purpose of an individual's life unfolds and becomes his "fate". Call it Sanjog or Synchronicity, it all happens through Divine Intervention.

***Our Collective Heritage:*** As children we are introduced to one of the many creeds of our Hindu religion-Sanatan Dharam, Jainism, Arya Samaj, Vedantism, Shivaism-or one of the many other creeds on the horizon. From then on we develop a religious complexion which stays with us for the rest of our lives. When we want more light, we read only one set of religious books to the total exclusion of others, lest by accident we get more light than we want.

A beam of white light passing through a prism splits up into a rainbow of colours. We select one of these colours, say the red one, and look at life under the red light. In our ignorance we refuse to look at life under the illumination of the blue light or the green light or the yellow light. We miss out on experiencing the true picture. Our shallow thinking prevents us from learning about the goodness in other creeds.

We are the descendents of the mighty Arjun, the sublime Krishan and the noble Ram. The teachings of Shankaracharya, Swami Dayanand, Guru Nanak, Buddh, Mahavir, Ramkrishan etc. are the collective heritage of all Hindus. There is no need to shut any one of them out. No matter what his religion is, the experience of every mystic is the same. And these mystics have left us with many powerful thoughts to ponder over and to illuminate our lives with.

As we celebrate India's Independence Day (or should we call it India's Freedom Day, as we never did depend on the colonial powers for our survival), let us celebrate our collective heritage and not be blinded by the monochromatic lights of different shades of Hinduism. Happy 15th folks!

***Conformity:*** From day one of his life, there is an attempt to shape every man into a common mold. The society, the school system and even the family demands conformity. But when you buckle under and become part of a herd, you lose your uniqueness. You give up your convictions and passions in favor of being accepted by the crowd. You become a mere actor playing different roles—role of a dutiful child, of a dutiful parent or of a dutiful spouse. Strive to find your own voice and do it now, because the longer you wait to begin the less likely you are to find it at all.

Psychoanalyst Erich Fromm wrote, "Man by origin is a herd animal. His actions are determined by an instinctive impulse to follow the leader and to have close contact with other animals around him. Inasmuch as we are sheep, there is no greater threat to our existence than to lose this contact with the herd and be isolated. But we are also human; we are endowed with reason which by its very nature is independent of the herd. Rationalization is a compromise between our sheep nature and our human capacity to think. And the full emergence of reason depends on our attaining independence so that our judgement is not based on our fear of being isolated from the herd. A few individuals can stand this isolation and say the truth in spite of the danger of losing touch. They are the true heroes of the human race but for whom we would still be living in caves."

Submission to a powerful authority gives a feeling of protection and belonging, but you lose your independence. Dare to strike out and find new ground. Be your own person. Look at things in a way different from the one you were trained to do. Don't be afraid of isolation. Resist the pressure to conform. If you don't, the casualty could be your heart and soul.

***On Flying a Kite:*** This is the kite flying season in Southern California. You can see the little ones in school parks struggling to get their kites up in the air. Erma Bombeck writes that kite flying is a lot like raising children. You spend what feels like a lifetime, getting them off the ground. You run with them until you are both breath less... they crash. You add a little longer tail and try again... they hit a tree. You rescue them, comfort them, fix them up with a couple of band-aids and try to get them up again. You watch them as they get lifted by the wind and finally become airborne. What a thrill it is to see them soar. Then they need more string and you keep unrolling your ball of twine. As the kite flies higher and higher, there is sadness mixed with joyful emotions. Deep inside you know that the beautiful kite of yours is going to get tangled with another kite and disappear into to the sunset. It will no longer need the lifeline that you so lovingly provided. And you feel the gratification by knowing that you did your job.

Too often when our children are little, we don't spend enough time with them. We are too busy doing our own things — having a busy social life, trying to make our first million or working to gain social prominence. We grow apart. Then one day, they are old enough to take care of themselves and don't need us anymore. We wake up, sometimes too late to enjoy them. And that is very sad. Because the joy that they provide, the vitality they infuse in their aging caretakers is overwhelming. They make us feel young again. Don't miss out on bonding with your adorable kites. Hold them and hug them and be part of their little games. Spend some quality time with the precious little angels. They are so much a reflection of you when you were little. You would have enough time to do your own things later on. After all they are going to be around only for about eighteen years before they get shipped off to some university.

***Mark Twain on 'What is Man':*** Man, according to Mark Twain is a mere machine. 'What so ever he is, is due to his make, and the influences brought to bear upon him by his heredities, his habitat and his associations.' Further more, 'From cradle to the grave a man never does a single thing which has any FIRST AND FOREMOST object but one-to secure peace of mind and spiritual comfort for HIMSELF.' Mark Twain claims that every man acts per his make. Men are born with some hereditary traits-gifts and shortcomings-and they act according to these traits. Just as a sewing machine made from different grades of steel or from different metals will perform differently, so would a tin man, a gold man or a steel man act according to the nature of his make. His make however is influenced by his training and his environment.

In one of his fables he says, 'A man is waiting at a bus station to reach home in the middle of a storm. A gray-haired ragged old homeless woman approaches him and requests for help. He ponders over for a few seconds and gives her his last dollar. He then walks home in the middle of a storm. On the way home his heart is singing with joy because he did something good. When he reaches home he proudly recounts his tale of charity and earns the respect and affection of his family. He has a hearty dinner and sleeps well that night, knowing how benevolent he has been to the old lady.

That is a very good return on an investment of one dollar.' Mark Twain concludes that earning his personal self-approval, securing his peace of mind and his desire to earn the affection of his family and to feel good about himself were the motivations that drove this man to help the old lady. Whatever he did was a result of his make, his training and above all an attempt to buy peace of mind for himself. There were no favors done. Noble, Generous, Benevolent etc. are merely undeserving labels that men bestow upon each other.

It sure is a different point of view, which makes you stop and think!

## Vedas and Philosophy

The Vedas are the most ancient books in the World, and they are the Foundation of Hinduism. Veda means knowledge. Any form of Knowledge acquired is considered as a Veda whereby it has no beginning or end. While it might surprise people how a book can have no beginning or end, the ancient Rishis who wrote these accepted that the complete knowledge of the Universe could never fit in any book, so there would always be new things to discover. This philosophy makes Hinduism a very tolerant religion, always ready to accept new ideas from other cultures.

The Vedas were compiled by the great sage Krishna Dwipayana during the Dwapara Yuga with the goal to come up with a de-facto standard of education. Upon gathering all the teachings passed on from the Acharyas (Teachers) to their Sishyas (Students) from Kingdom to Kingdom, he compiled them into 4 standard structures; the Rigveda, Yajurveda, Samaveda and Atharvaveda. Upon completion of this great feat he was given the title "Veda Vyasa" which means "Compiler of the Vedas".

The Vedas are considered Shruti (or Sruti), or revealed texts. They were not given by a prophet, but heard by many different Rishis (or very advanced Yogis) during deep meditation. These verses were combined and written in the Vedas in poetic form.

The Vedas are not Polytheistic. Dr. David Frawley, in his book "Wisdom of the Ancient Seers" mentions "The Gods, though they have a human facet, are not anthropomorphic. The Gods represent not the Divine in the image of Man, but rather man in the image of Divine, in the image of all creation."

Regarding the charge of Pantheism, he says in the same book "The natural imagery of the ancients reflects not the Divine reduced to natural world but the natural world as a reflection of the inner truth".

So Agni, the God of Fire, is not normal fire, but the fire of transformation. He burns our inner demons, and lights the way to Enlightenment. Soma, the God of Wine, is the Divine bliss we feel when reach the Divine, and feel him in all creation. Saraswati is not a river or Goddess of a River, but rather the River of Divine knowledge that continuously flows from heaven. When the Demons block this river and the Gods fight them, it is not a literal battle, but a figurative one when the Divine qualities inherent in Man fight the Demons of ignorance.

***Organization:*** The Mantras are collected into anthologies called Samhitas. There are four Samhitas: the *Rk* (poetry), *Sman* (song), *Yajus* (prayer), and *Atharvan* (a kind of priest). They are commonly referred to as the Rigveda, Samaveda, Yajurveda, and Atharvaveda respectively. Each Samhita is preserved in a number of versions or recensions (shakhas), the differences among them being minor, except in the case of the Yajur Veda, where two "White" (*shukla*) recensions contain the Mantras only, while four "Black" (*krishna*) recensions interspersed the Brahmana parts among the Mantras.

The Rigveda contains the oldest part of the corpus, and consists of 1028 hymns. The Samaveda is mostly a rearrangement of the Rigveda for musical rendering. The Yajurveda gives sacrificial prayers and the Atharvaveda gives charms, incantations and magical formulae. In addition to these there are some stray secular material, such as legends.

The next category of texts are the Brahmanas. These are ritual texts that describe in detail the sacrifices in which the Mantras were to be used, as well as comment on the meaning of the sacrificial ritual. Each of the Brahmanas is associated with one of the Samhitas. The Brahmanas may either form separate texts, or in the case of the Black Yajur Veda, can be partly integrated into the text of the Samhita. The most important of the Brahmanas is the Shatapatha Brahmana of the White Yajur Veda.

The Aranyakas and Upanishads are theological and philosophical works. They are mystic or spiritual interpretations of the Vedas, and are considered their putative end and essence, and thus known as Vedanta ("the end of the Vedas"). They often form part of the Brahmanas (*e.g.* the Brhadaranyaka Upanishad). They are the basis of the Vedanta school of Darsana.

***Position and Compilation:*** Hindu tradition regards the Vedas as uncreated, eternal and being revealed to sages (Rishis). The rishi Krishna Dwaipayana, better known as Veda Vyasa – "Vyasa" meaning "editor" or "compiler" – reputedly distributed this mass of hymns into the four books of the Vedas, each book being supervised by one of his disciples.

Paila arranged the hymns of the *Rig Veda*. Those that were chanted during religious and social ceremonies were compiled by Vaishampayana under the title *Yajus mantra Samhita* (see Yajur-Veda). Jaimini is said to have collected hymns that were set to music and melody— "Saman" (see Sama-Veda). The fourth collection of hymns and chants known as the *Atharva Samhita* was collated by Sumanta.

Philosophies and sects that developed in the Indian subcontinent have taken differing positions on the Vedas. In Buddhism and Jainism, the authority of the Veda is repudiated, and both evolved into separate religions. The sects which did not explicitly reject the Vedas remained followers of the Sanatana Dharma, which is known in modern times as Hinduism.

***Study:*** Elaborate methods for preserving the text (memorizing by heart instead of writing), subsidiary disciplines (Vedanga), exegetical literature, etc., were developed in the Vedic schools. Sayana, from the 14th century, is known for his elaborate commentaries on the Vedic texts. While much evidence suggests that everyone was equally allowed to study the Vedas and many Vedic "authors" were women, the later dharmashastras, from the Sutra age, dictate that women and Shudras were neither required nor allowed to study the Veda. These dharmashastras regard the study of the Vedas a religious duty of the three upper varnas (Brahmins, Kshatriyas and Vaishyas). In modern times, Vedic studies are crucial in the understanding of Indo-European linguistics, as well as ancient Indian history.

Many forms of Hinduism encourage the Vedic mantras to be interpreted as liberally and as philosophically as possible, unlike the texts of the three Abrahamic religions. In fact, over-literal interpretation of the mantras is actually discouraged, and even the three layers of commentaries (Brahmanas, Aranyakas, and Upanishads), which form an integral part of the Sruti literature, interpret the seemingly polytheistic, ritualistic, and highly complex Samhitas in a philosophical and metaphorical way to explain the "hidden" concepts of God (Ishwara), the Supreme Being (Brahman) and the soul or the self (Atman). Many Hindus believe that the very sound of the Vedic mantras is purifying for the environment and the human mind.

***Cosmogony:*** The Vedic view of the world and cosmogony sees one true divine principle self-projecting as the divine word, *Vaak*, 'birthing' the cosmos that we know from 'Hiranyagarbha' or Golden Womb, a primordial sun figure that is equivalent to Surya. The varied gods like Vayu, Indra, Rudra (the Destroyer), Agni (Fire, the sacrificial medium) and the goddess Saraswati (the Divine Word, aka Vaak) are just some examples of the myriad aspects of the one underlying nature of the universe.

***Dating:*** Many historians regard the Vedas as one of the oldest surviving texts (see the ancient Egyptian texts "The

Story of Sinuhe" and the "Ipuwer Papyrus", both dated 1800 BCE). The newest parts of the Vedas are estimated to date to around 500 BCE; the oldest text (Rig Veda) is found to have been completed by 1500 BCE after being composed over hundreds of years, but most Indologists agree that a long oral tradition possibly existed before it was written down.

They represent the oldest stratum of Indian literature and according to modern scholars are written in forms of a language which evolved into Sanskrit. They consider the use of Vedic Sanskrit for the language of the texts an anachronism, although it is generally accepted. Some writers have used astronomical references in the Rigveda to date it to as early as the 4th millennium BC.

The Hindus fundamental belief is that the Vedas are sanatan-eternal-and apaurusheya not composed by human entity. At the beginning of every cosmic cycle of brahma, Paramatma utters the divine words. Later, at various periods great rishis perceive these divine words and imparted this knowledge orally through generations. It is believed that this Vedic knowledge totally disappeared at the end of Brahmas cosmic cycle and reappeared again in Brahma's next cycle of creation. Later a part of vedic knowledge was written. Maha Rishi Veda Vyas simplified this one Veda by dividing it into four: Rig, Sam, Yajur, Atharva. In Bharatvarshas true tradition the vedas are eternal.

***Etymology:*** The word "veda" means "knowledge", and is derived from the root "vid", Sanskrit for "know", reconstructed as being derived from the Proto-Indo-European root "weid-", meaning "see" or "know". "Weid-" is also the source of the English word "wit", as well as "vision" through Latin. The Czech and Slovak words for "science" are "vda" resp. "veda", derived from western Slavic "vdet" resp. "vediee" for "know".

## Derivation from the Vedâs-Upavedas

The Upavedas are derived from the Vedas and are specific applications of the teachings of the Vedas. The main Upavedas are:

1. *Ayurveda:* Indias healing system, it lays more stress on living with nature instead of fighting it, hence preventive instead of corrective medicine.
2. *Dhanur Veda:* Martial arts.

   Ayurveda and Dhanurveda have points in common. They both work with Marma, or natural Pran (Life Energy) that flows in the body.

   Ayruveda heals the body, while Dhanurveda is used for killing. This concept is also known to Chinese as Acupuncture and related Chinese Martial Arts.
3. *Stahapatya Veda:* Architecture, sculpture and geomancy. Used especially for Temple design.
4. *Gandharv Veda:* Music, poetry and dance.

Some other fields like Jyotish (Indian Astrology), Tantra (based on the Puranas, which are in turn based on Vedas), Shiksha and Vyakara (Grammar and pronunciation) are also based on the Vedas.

The six schools of Vedic Philosophy

1. Nyaya—The Logical School, founded by Gautama
2. Vaishesika—Atomic school, founded by Kannada
3. Samkhya—Cosmic Principle School, founded by Kapila
4. Yoga—Yoga school (includes Raj, Hatha and Tantra Yoga), founded by Hiranyagarbha, although Patanjalis Sutras are the most popular book that has survived.
5. Purva Mimamsa—Ritualistic School, founded by Jamini
6. Uttara Mimasa/Vedant—Theological School, founded by Badarayana.

Vedanta was made popular by Adi Shankara also called Shankaracharya, who founded the Swami Order of Monks, and established 4 schools (or Maths) in 4 parts of India to carry on teachings of the Vedas in the 7th century.

***The Hidden Meaning of the Vedas:*** The Vedas were written in poetic language. By literally translating them, Western translators lost some of the poetic beauty. Their interpretations of the shastras are often shallow. Plus, they had to force meanings where there were none. Words like

Pantheism, Polytheism are used to describe Hinduism, but this shows the translators ignorance or bias. One of the best commentaries to Vedas is written by Sri Aurobindo. Rig Veda is considered by many to be a book written by barbaric culture worshipping violent Gods. Aurobindo realized that this was due to the biased view of Westerners who had some preconceived views on Hindu culture.

So Aurobindo decided to look for hidden meanings in the Vedas. He looked at the Rig Veda as a psychological book, inspiring the people to move towards God, but in a hidden language. So Indra is the God of Indriya, or the senses (sight, touch, hear, taste etc.). Varun means air, but in esoteric terms means Pran, or the Life force. So when the Rig Vedas says "Call Indra and Varun to drink Soma Rasa" they mean use the Mind senses and Pran to receive divine bliss (Soma means wine of Gods, but in several texts also means Divine Bliss, as in Right handed Tantra).

Agni, or God of Fire, is the hidden Divine Spark in us, which we have to fan, so it grows and engulfs our whole body. So the sacrifice of the Vedas actually means sacrificing ones ego to the internal Agni, or Divine spark. These essays originally appeared in the Arya, but have been condensed as a book form as "The Secret of the Vedas" by Sri Aurobindo.

***Westerners and Vedas:*** The Vedas are hidden in mystic language. The Rishis hid 3-4 esoteric meanings within each verse, and it required a good Guru to explain them.

One of the few Westerners to study the Vedas with an Indian Guru is Dr. David Frawley. In his excellent book "Wisdom of the Ancient Seers" he explores ideas similar to one by Sri Aurobindo. He describes Indra as a brave soul, willing to fight evil and darkness. He is the inner Spirit, wanting to break free from the fake shackles of ego.

Frawley, who also studied Ayurveda and Tantra, says Agni also refers to Kundalini, and Soma Rasa to the Sahasra Chakra and the bliss that flows from it. Savitur is the Sun that shines in the darkness of ignorance, and guides us towards the Truth.

## Samhita

The Samhita (Sanskrit: "joined" or "collected") is the basic text of each of the Vedas, comprising collections of hymns and ritual texts. This term was originally used in reference to the style of recitation used during hymns and chants. In the Vedas the Samhitas are supplemented by later explanatory commentaries, notably the Brahmanas and Upanishads.

The term is also sometimes spelled 'Sanhita'.

***Other Samhitas:*** In later times, the term Samhita also came to be applied to the names of texts belonging to various fields of knowledge, not directly connected to the Vedas, *e.g.*

- Gherand Samhita
- Charak Samhita
- Ashtavakra Samhita

***Gherand Samhita:*** Gheranda Samhita (Sanskrit for "Gheranda's Collection") is one of the three classic texts of Hatha Yoga (the other two being the Hatha Yoga Pradapika and the Shiva Samhita). It is a late 17th Century text and is considered to be the most encyclopedic of the three classic texts on Hatha Yoga.

Gheranda Samhita is a manual of Yoga taught by Gheranda to Chanda. Unlike other Hatha Yoga texts, the Gheranda Samhita focuses upon the *Shat Kriyas* or internal body cleansing, *i.e.*, *Ghatastha Yoga*. The closing stanzas of the Gheranda Samhita deal with Samadhi, but the methods taught are different from Patanjali's methods.

***Charaka:*** Charaka, sometimes spelled Caraka, (300 BC) is one of the founders of Ayurveda. According to Charaka, health and disease are not predetermined and life may be prolonged by human effort.

***Contributions:*** According to the Charaka tradition, there existed six schools of medicine, founded by the disciples of the sage Punarvasutreya. Each of his disciples, Agnivesha, Bhela, Jatkkarna, Parshara, Hrita, and Kshrapni, composed a Samhit. Of these, the one composed by Agnivesha was supposed to be the best. The Agnivesha Samhit was later revised by Charaka and it came to be known as Charaka Samhit. The Charaka Samhit was revised by Dridhbala.

Yurveda is traditionally divided into eight branches which, in Charaka's scheme, are:

1. sktra-sthna, general principles
2. nidna-sthna, pathology
3. vimna-sthna, diagnostics
4. shamrasthna, physiology and anatomy
5. indriya-sthna, prognosis
6. chikits-sthna, therapeutics
7. kalpa-sthna, pharmacy
8. siddhi-sthna, successful treatment

***Form of Charak Samhita:*** Charak Samhita contains 120 adhyayas (chapters) in all. These are devided in 8 parts:

1. Sutra sthana 30
2. Nidan sthana 08
3. Viman sthana 08
4. sharir sthana 08
5. Indriya sthana 12
6. Chikitsa sthana 30
7. Kalpa sthana 12
8. Siddhi sthana 12

## Ashtavakra Gita

The Ashtavakra Gita (Song of Ashtavakra), also known by the name Ashtavakra Samhita is an influential non-dualist Vedic scripture traditionally said to have been spoken by the Sage Ashtavakra, though its authorship is not known with certainty. There is little doubt though that it is very old, probably dating back to the days of the classic Vedanta period. The Sanskrit style and the doctrine expressed would seem to warrant this assessment. The work was known, appreciated and quoted by Ramakrishna and his disciple Vivekananda, as well as by Ramana Maharshi. Radhakrishnan refers to it with great respect, while even Sri Ravi Shankar has given a commentary on the scripture. Apart from that, the work speaks for itself. It presents the traditional teachings of Advaita Vedanta with a clarity and power very rarely matched.

The *Ashtavakra Gita* does not date itself or brand itself to any region, culture, or peoples, although it does contain Hindu-specific references, especially in the final chapter entitled 'I am Shiva' where it is proclaimed in the final verse that the author is in truth one with God. The essence of the *Ashtavakra Gita* is that there is no such thing as existence or non existence, right or wrong, or moral or immoral. In the eyes of the sage Ashtavakra who is the ostensible author of this text, one's true identity can be found by simply recognizing oneself as Pure Existence and that as individuals we are the Awareness of all things.

The Ashtavakra Gita teaches that one is already free once one realises they are free. It advocates non-action (similar to the Daoist concept of Wu Wei), the loss of desire and severing of worldly attachments. To free oneself from the cycle of life and death one should withdraw from all Earthly desires, worries and cares. To continue including in Earthly things even after one has realized their true nature is said to be foolish and time wasting. Instead it paints a picture of The Master as someone who continues to keep up their responsibilities in the world, not because they believe they have to or due to any worldy attachments, but simply that it is in their nature to do so. It is important to avoid misinterpretation in this regard, and to that end teachers traditionally recommend that Ashtavakra Gita be pursued by only those who have already advanced on the spiritual path. Ashtavakra teaches that emotion, thought and even meditation can lead to further bondage and that religious practice itself is of no meaning to the enlightened.

## Rigveda

The Rigveda consists of 1,028 hymns (or 1,017 discounting the apocryphal *valakhalya* hymns 8.49–8.59) composed in Vedic Sanskrit, many of which are intended for various sacrificial rituals. This long collection of short hymns is mostly devoted to the praise of the gods. It is organized into 10 books, known as Mandalas. Each mandala is comprised of hymns, called *skkta*, which are further comprised of individual. The Mandalas are by no means of equal length or age: The "family books", mandalas 2-7, are considered the oldest part and comprise the shortest books, arranged by length, accounting for 38% of the text. RV 8 and RV 9, likely comprising hymns of mixed age, account for 15% and 9%, respectively. RV 1 and RV 10, finally, are both the youngest and the longest books, accounting for 37% of the text.

***Preservation:*** The *Rigveda* is preserved by two major shakhas ("branches", *i.e.* schools or recensions), *Zkala* and *Bckala*. Considering its great age, the text is spectacularly well preserved and uncorrupted, the two recensions being practically identical, so that scholarly editions can mostly do without a critical apparatus. Associated to *Zkala* is the Aitareya-Brahmana.

The *Bckala* includes the Khilani and has the Kausitaki-Brahmana associated to it. This compilation or redaction included the arrangement in books as well as orthoepic changes, such as regularization of sandhi (called by Oldenberg *orthoepische Diaskeunase*). It took place centuries after the composition of the earliest hymns, about coeval to the redaction of the other Vedas.

From the time of its redaction, the text has been handed down in two versions: The *Samhitapatha* has all Sanskrit rules of sandhi applied and is the text used for recitation. The Padapatha has each word isolated in its pausa form and is used for memorization. The Padapatha is, as it were, a commentary to the *Samhitapatha*, but the two seem to be about coeval. The original text as reconstructed on metrical grounds (*viz.* "original" in the sense that it aims to recover the hymns as composed by the Rishis) lies somewhere between the two, but closer to the Samhitapatha.

***Organization:*** Hermann Grassmann has though numbered the hymns 1 through to 1028, putting the *valakhilya* at the end. The entire 1028 hymns of the *Rigveda*, in the 1877 edition of Aufrecht, contain a total of 10,552 verses, or 39,831 padas. The

Shatapatha Brahmana gives the number of syllables to be 432,000, while the metrical text of van Nooten and Holland (1994) has a total of 395,563 syllables (or an average of 9.93 syllables per pada); counting the number of syllables is not straightforward because of issues with sandhi. Most verses are jagati (padas of 12 syllables), trishtubh (padas of 11 syllables), viraj (padas of 10 syllables) or gayatri or anushtubh (padas of 8 syllables).

The chief gods of the *Rigveda* are Agni, the sacrificial fire, Indra, a heroic god who is praised for having slain his enemy Vrtra, and Soma, the sacred potion, or the plant it is made from. Other prominent gods are Mitra, Varuna and Ushas (the dawn) and the Ashvins.

Also invoked are Savitar, Vishnu, Rudra, Pushan, Brihaspati, Brahmanaspati, Dyaus Pita (the sky), Prithivi (the earth), Surya (the sun), Vayu (the wind), Apas (the waters), Parjanya (the rain), Vac (the word), the Maruts, the Adityas, the Rebus, the Vishvadevas (the all-gods), many rivers (notably the Sapta Sindhu, and the Sarasvati River), as well as various further minor gods, persons, concepts, phenomena and items. It also contains fragmentary references to possible historical events, notably the struggle between the early Vedic people (known as Vedic Aryans, a subgroup of the Indo-Aryans) and their enemies, the Dasa.

*Rigveda manuscript in Devanagari, early 19th century*

- Mandala 1 comprises 191 hymns. Hymn 1.1 is addressed to Agni, and his name is the first word of the *Rigveda*. The remaining hymns are mainly addressed to Agni and Indra. Hymns 1.154 to 1.156 are addressed to Vishnu.
- Mandala 2 comprises 43 hymns, mainly to Agni and Indra. It is chiefly attributed to the Rishi *gitsamda saunohotra*.
- Mandala 3 comprises 62 hymns, mainly to Agni and Indra. The verse 3.62.10 has great importance in Hinduism as the Gayatri Mantra. Most hymns in this book are attributed to *visvmitra gthinas*.

- Mandala 4 consists of 58 hymns, mainly to Agni and Indra. Most hymns in this book are attributed to *vmadeva gautama*.
- Mandala 5 comprises 87 hymns, mainly to Agni and Indra, the Visvadevas, the Maruts, the twin-deity Mitra-Varuna and the Asvins. Two hymns each are dedicated to Ushas (the dawn) and to Savitar. Most hymns in this book are attributed to the *atri* family.
- Mandala 6 comprises 75 hymns, mainly to Agni and Indra. Most hymns in this book are attributed to the *brhaspatya* family of Angirasas.
- Mandala 7 comprises 104 hymns, to Agni, Indra, the Visvadevas, the Maruts, Mitra-Varuna, the Asvins, Ushas, Indra-Varuna, Varuna, Vayu (the wind), two each to Sarasvati and Vishnu, and to others. Most hymns in this book are attributed to *vasicmha maitravaurni*.
- Mandala 8 comprises 103 hymns to different gods. Hymns 8.49 to 8.59 are the apocryphal *valakhalya*. Most hymns in this book are attributed to the *kyva* family.
- Mandala 9 comprises 114 hymns, entirely devoted to Soma *Pavamana*, the plant of the sacred potion of the Vedic religion.
- Mandala 10 comprises 191 hymns, to Agni and other gods. It contains the Nadistuti sukta which is in praise of rivers and is important for the reconstruction of the geography of the Vedic civilization and the Purusha sukta which has significance in Hindu tradition. It also contains the Nasadiya sukta (10.129), probably the most celebrated hymns in the west, which deals with creation.

***Rishis:*** Each hymn of the Rigveda is traditionally attributed to a specific rishi, and the "family books" (2-7) are said to have been composed by one family of rishis each. The main families, listed by the number of verses ascribed to them are:

- Angirasas: 3619 (especially Mandala 6)
- Kanvas: 1315 (especially Mandala 8)

- Vasishthas: 1267 (Mandala 7)
- Vaishvamitras: 983 (Mandala 3)
- Atris: 885 (Mandala 5)
- Bhrgus: 473
- Kashyapas: 415 (part of Mandala 9)
- Grtsamadas: 401 (Mandala 2)
- Agastyas: 316
- Bharatas: 170

The *Rigveda* was translated into English by Ralph T.H. Griffith in 1896. Partial English translations by Maurice Bloomfield and William Dwight Whitney exist. Griffith's translation is good, considering its age, but it is no replacement for Geldner's 1951 translation (in German), the only independent scholarly translation so far. The later translations by Elizarenkova depends heavily on Geldner, but Elizarenkova's translation (in Russian) is valuable in taking into account scholarly literature up to 1990.

***Hindu Tradition:*** According to Indian tradition, the Rigvedic hymns were collected by Paila under the guidance of Vysa, who formed the Rigveda Samhita as we know it. According to the Zatapatha Brhmana, the number of syllables in the *Rigveda* is 432,000, equalling the number of muhurtas (1 day = 30 muhurtas) in forty years. This statement stresses the underlying philosophy of the Vedic books that there is a connection (bandhu) between the astronomical, the physiological, and the spiritual.

The authors of the Brhmana literature described and interpreted the Rigvedic ritual. Yaska was an early commentator of the *Rigveda*. In the 14th century, Syana wrote an exhaustive commentary on it. Other *Bhacyas* (commentaries) that have been preserved up to present times are those by Madhava, Skacdasvmin and Veckatamdhava.

***Dating and Historical Reconstruction:*** Geography of the Rigveda, with river names; the extent of the Swat and Cemetery H cultures are also indicated. The *Rigveda* is far more archaic than any other Indo-Aryan text. For this reason, it was in the center of attention of western scholarship from the times of Max Muller. The *Rigveda* records an early stage of Vedic religion, still closely tied to the pre-Zoroastrian Persian religion. It is thought that Zoroastrianism and Vedic Hinduism evolved from an earlier common religious Indo-Iranian culture.

The Rigveda's core is accepted to date to the late Bronze Age, making it the only example of Bronze Age literature with an unbroken tradition. Its composition is usually dated to roughly between 1700–1100 BC The text in the following centuries underwent pronunciation revisions and standardization (samhitapatha, padapatha). This redaction would have been completed in about the 7th century BC Writing appears in India in ca. the 5th century BC in the form of the Brahmi script, but texts of the length of the Rigveda were likely not written down before the Early Middle Ages, in the Gupta or Siddham scripts, and while written manuscripts were used for teaching in medieval times, they played a minor role in the preservation of knowledge because of their ephemeral nature (Indian manuscripts were on bark or palm leaves and decomposed rapidly in the tropical climate) until the advent of the printing press in British India.

The hymns were thus preserved by oral tradition for up to a millennium from the time of their composition until the redaction of the Rigveda, and the entire Rigveda was preserved in shakhas for another 2,500 years from the time of its redaction until the *editio princeps* by Muller, a collective feat of memorization unparalleled in any other known society.

Puranic literature names Vidagdha as the author of the Pada-text. Other scholars argue that Sthavira Sak of the Aitareya Aranyaka is the padakara of the RV. After their composition, the texts were preserved and codified by a vast body of Vedic priesthood as the central philosophy of the Iron Age Vedic civilization.

The *Rigveda* describes a mobile, nomadic culture, with horse-drawn chariots and metal (bronze) weapons. According to some scholars the geography described is consistent with that of the Punjab (Gandhara): Rivers flow north to south, the mountains are relatively remote but still reachable (Soma is a plant found in the mountains, and it has to be purchased, imported by merchants). Nevertheless, the hymns were certainly composed over a long period, with the oldest elements possibly reaching back into Indo-Iranian times, or the early 2nd millennium BC.

Thus there is some debate over whether the boasts of the destruction of stone forts by the Vedic Aryans and particularly by Indra refer to cities of the Indus Valley civilization or whether they hark back to clashes between the early Indo-Aryans with the BMAC (Bactria-Margiana Archaeological Complex) culture centuries earlier, in what is now northern Afghanistan and southern Turkmenistan (separated from the upper Indus by the Hindu Kush mountain range, and some 400 km distant). In any case, while it is highly likely that the bulk of the *Rigveda* was composed in the Punjab, even if based on earlier poetic traditions, there is no mention of either tigers or rice in the *Rigveda* (as opposed to the later Vedas), suggesting that Vedic culture only penetrated into the plains of India after its completion.

Similarly, it is assumed that there is no mention of iron. The Iron Age in northern India begins in the 12th century BC with the Black and Red Ware (BRW) culture. This is a widely accepted timeframe for the beginning codification of the *Rigveda* (*i.e.* the arrangement of the individual hymns in books, and the fixing of the samhitapatha (by applying Sandhi) and the padapatha (by dissolving Sandhi) out of the earlier metrical text), and the composition of the younger Vedas. This time probably coincides with the early Kuru kingdom, shifting the center of Vedic culture east from the Punjab into what is now Uttar Pradesh.

Some of the names of gods and goddesses found in the *Rigveda* are found amongst other belief systems based on Proto-Indo-European religion as well: Dyaus-Pita is cognate with Greek Zeus, Latin Jupiter (from deuspater), and Germanic Tyr; while Mitra is cognate with Persian *Mithra*; also, Ushas with Greek Eos and Latin Aurora; and, less certainly, Varuna with

Greek Uranos. Finally, Agni is cognate with Latin *ignis* and Russian *ogon*, both meaning "fire".

Some writers have traced astronomical references in the *Rigveda* dating it to as early as 4000 BC· a date well within the Indian Neolithic. Claims of such evidence remain controversial. Kazanas (2000) in a polemic against the "Aryan Invasion Theory" suggests a date as early as 3100 BC, based on an identification of the early Rigvedic Sarasvati River as the Ghaggar-Hakra and on glottochronological arguments. Being a polemic against mainstream scholarship, this is of course in diametral opposition to views in mainstream historical linguistics, which assumes a date as late as 3000 BC for the age of late Proto-Indo-European itself.

***Flora and Fauna in the Rigveda:*** The horse (Asva) and cattle play an important role in the Rigveda. There are also references to the elephant (Hastin, Varana), Camel (Ustra, especially in Mandala 8), Buffalo (Mahisa), lion (Simha) and to the Gaur in the Rigveda. The peafowl (Mayura) and the Chakravaka (Anas casarca) are birds mentioned in the Rigveda.

***More Recent Indian Views:*** The Hindu perception of the *Rigveda* has moved away from the original ritualistic content to a more symbolic or mystical interpretation. For example, instances of animal sacrifice are not seen as literal slaughtering but as transcendental processes. The Rigvedic view is seen to consider the universe to be infinite in size, dividing knowledge into two categories: *lower* (related to objects, beset with paradoxes) and *higher* (related to the perceiving subject, free of paradoxes). Swami Dayananda, who started the Arya Samaj and Sri Aurobindo have emphasized a spiritual (adhyatimic) interpretation of the book.

The Sarasvati river, lauded in RV 7.95 as the greatest river flowing from the mountain to the sea is sometimes equated with the Ghaggar-Hakra river, which went dry perhaps before 2600 BC or certainly before 1900 BC. Others argue that the Sarasvati was originally the Helmand in Afghanistan. These questions are tied to the debate about the Indo-Aryan migration (termed "Aryan Invasion Theory") vs. the claim that Vedic culture together with Vedic Sanskrit originated in the Indus Valley Civilization, a topic of great significance in Hindu nationalism, addressed for example by Amal Kiran and Shrikant G. Talageri. Subhash Kak has claimed that there is an astronomical code in the organization of the hymns. Bal Gangadhar Tilak, also based on astronomical alignments in the Rigveda, in his "The Orion" (1893) claimed presence of the Rigvedic culture in India in the 4th millennium BC, and in his "Arctic Home in the Vedas" (1903) even argued that the Aryans originated near the North Pole and came south during the Ice Age.

## Samaveda

The Samaveda is third in the usual order of enumeration of the four Vedas, the ancient core Hindu scriptures. The Samaveda ranks next in sanctity and liturgical importance to the Rigveda or Veda of Recited praise. Its Sanhita, or metrical portion, consists chiefly of hymns to be chanted by the Udgatar priests at the performance of those important sacrifices in which the juice of the Soma plant, clarified and mixed with milk and other ingredients, was offered in libation to various deities. The Sama Veda was compiled around 1300-1000 BC.

The Collection is made up of hymns, portions of hymns, and detached verses, taken mainly from the Rigveda, transposed and rearranged, without reference to their original order, to suit the religious ceremonies in which they were to be employed. The verses are not intended to be chanted, but to be sung in specifically indicated melodies using the seven *svaras* or notes. Such songs are called Samagana and in this sense the Samaveda is really a book of hymns.

In these compiled hymns there are frequent variations, of more or less importance, from the text of the Rigveda as we now possess it which variations, although in some cases they are apparently explanatory, seem in others to be older and more original than the readings of the Rigveda. In singing, the verses are still further altered by prolongation, repetition and insertion of syllables, and various modulations, rests, and other modifications prescribed, for the guidance of the officiating priests, in the Ganas or Song-books. Two of these manuals, the Gramageyagana, or Congregational, and the Aranyagana or Forest Song-Book, follow the order of the verses of part I, of the Sanhita, and two others, the Uhagana, the Uhyagana, of Part II. This part is less disjointed than part I, and is generally arranged in triplets whose first verse is often the repetition of a verse that has occurred in part I.

The Samaveda survives in a single shakha or recension, the Kauthuma shakha, with a second shakha, Jaiminiya (or Talavakara), surviving fragmentarily, the Jaiminiya Samhita. From the Jaiminiya shakha, we also have the Jaiminiya Brahmana, the Jaiminiya Upanishad Brahmana and the Kena Upanishad.

## Yajurveda

The Yajurveda is one of the four Hindu Vedas. It contains religious texts focusing on liturgy, rituals and sacrifices, and how to perform the same. The Yajurveda was probably written around 900 BCE.

***Collections:*** There are two primary collections or *samhitas* of the Yajurveda: Shukla (white) and Krishna (black). Both contain the verses necessary for rituals, but the Krishna Yajurveda has additional prose commentary and detailed instructions within the work itself.

***Krishna Yajurveda:*** There are four recensions of the Krishna ("black" or "dark") Yajurveda:

- *taittiraya sachita* (TS) of Panchala
- *maitrayani sachita* (MS)
- *caraka-katha sachita* (KS) of Madra
- *kapicmhala-katha sachita* (KapS) of Bahika

Each of the recensions has a Brahmana associated with it, and some of them also have associated Shrautasutras, Grhyasutras, Aranyakas, Upanishads and Pratishakhyas.

The best known of these recensions is the TS, named after Tittiri, a pupil of Yaska. It consists of 8 books or *kandas*, subdivided in chapters or *prapathakas*, further subdivided into individual hymns. Some individual hymns in this Samhita have gained particular importance in Hinduism; *e.g.* TS 4.5 and TS 4.7 constitute the Shri Rudram Chamakam, while 1.8.6.1 is the Shaivaite Tryambakam mantra. The formula *bhkr bhuvas, suvas* prefixed to the (rigvedic) Gayatri mantra is also from the Yajurveda, appearing four times.

***The Taittiriya Shakha:*** The Taittiriya Shakha (or branch) of the Krishna Yajur Veda is the *shakha* most prevalent in southern India. Among the followers of this Shakha, the Apastamba Sutras is the common Shrautasutra associated with the Shakha. The Taittiriya Shakha consists of Taittiriya Samhita (having seven kaandas), Taittiriya Braahmana (having three kaandas), Taittiriya Aaranyaka (having seven prashnas) (See Aranyaka Literature), Taittiriya Upanishad (having three *prashnas* or *vallis*-Sheeksha valli, Ananda valli and Bhrigu valli) and the Mahaanaarayana Upanishad.

The Taittiriya Upanishad and Mahaanaarayana Upanishad are considered to be the seventh, eighth, ninth and tenth prashnas of the Aaranyaka. The words *prapaathaka* and *kaanda* (meaning sections) are interchangeably used in the Vedic literature. *Prashna* and *valli* refer to sections of the Aaranyaka.

***Shukla Yajurveda:*** There are two shakhas or recensions of the Shukla ("white" or "bright") Yajurveda:

- *Vajasaneyi Madhyandiniya*
- *Vajasaneyi Kanva* of Kosala

The former is popular in North India and Orissa, and thus commands a numerous following. The Kanva Shakha is popular in Karnataka, Kerala Andhrapradesh and Tamil Nadu. Jagadguru Adi Shankara is said to have followed the Kanva *Shakha* (recension). The Vedic rituals of the Ranganathaswamy Temple at Srirangam, the biggest temple in India, are performed according to the latter *Shakha*. Kanva matha the only mutt formed on the basis of Veda known as Veda matha was established by saint Madhava thirtha during 1898 who got sanyasa ashrama by sri lakshmi manohara thirtha of sirur math of Udupi one of the ashtamata established by sage Madhvacharya to conduct pooja kainkaryas to lord Udupi

Krishna the Kanva matha has its headquarter at Hunasihole at Gulbarga district of Karnataka state with sri sri Vidyabhaskara thirtha swamiji as its head now.

The Shukla Yajur Veda has two Upanishads associated with it: the Isa Vasya and the Brihadaranyaka upanishads. The Brihadaranyaka Upanishad is the most voluminous of all Upanishads.

***Legend:*** Legend has it that the vedic seer Yajnavalkya studied the Yajurveda collections under the tutelage of sage Vaishampayana maternal uncle of Yajnavalkya. Yajnavalkya's birth was with a purpose as purported by Gods. He was an 'Ekasandhigraahi' meaning he learns with just once teaching it self, he was too intelligent. The two came to have serious differences in interpretation. On one occasion, Vaishampayana was so enraged that he demanded the return of all the knowledge he has imparted to Yajnavalkya. Yagnavalkya regurgitated all the knowledge he had learnt in the form of flesh. The other disciples of Vaishampayana assumed the form of *tittiriya* birds and ate that flesh, eager to receive the knowledge.

Thus, that knowledge came to be called the Taittiriya Samhita (from the word *tittiriya*). There is a very secret meaning in this. As we know; it is impossible to regurgiate a knowledge or learning. But the words are used in 'srimad bhaagavata'. It is also of interest to know why the Guru him self did not taught the vedas he tought to yaajnyavalkya instead asked his disciples to eat the feelings. Any way, after having regurgitated out the knowledge acquired from his teacher, Yajnavalkya worshipped Surya (the Sun God) and acquired new knowledge directly from God Narayana (suryantargata narayana) who preached Shukla yajurveda by taking vaaji rupa, (god with horse face) which was compiled to become the *Shukla Yajurveda, this* is recorded in srimad bhaagavata. (haya, vaaji, ashwa, turaga are names referring horse in different lokaas)

## Atharvaveda

The Atharvaveda is a sacred text of Hinduism, and one of the four Vedas, often called the "fourth Veda". According to tradition, the Atharvaveda was mainly composed by two groups of rishis known as the Bhrigus and the Angirasas. Additionally, traditions ascribes parts to other rishis, such as Kauyka, Vaycmha and Kacyapa. There are two surviving recensions (lkhs), known as Zaunakiya (AVS) and Paippalda (AVP). The fixation of the samhita texts of these recensions likely dates to roughly 600 BCE (the "Mantra" period of Vedic Sanskrit, and the following codification in Vedic shakhas).

***Status:*** The Atharvaveda, while undoubtedly belonging to the core Vedic corpus, in some ways represents an independent parallel tradition to that of the Rigveda and Yajurveda. The Jaina and Buddha texts are considerably more hostile to the AV (they call it Aggvna or Ahavna Veda) than they are to the other Hindu texts. They even call it a non-Aryan Veda concocted by paippalda for human sacrifices. The Hindu texts too have taken a less than charitable view and have on occasions omitted the reference to the "Atharvn" text in the context of Vedic literature, though some attribute this to the fact that the Atharvaveda was a later addition chronologically. The Atharvan Pariththas (appendices) themselves state that specific priests of the Mauda and Jalada schools should be avoided. It is even stated that women associated with Atharvn may suffer from abortions.

***Recensions:*** The *Cara Gavyuha* (attributed to Shaunaka) lists nine shakhas or Schools of the Atharvaveda:

1. paippalda
2. stauda
3. mauda
4. aunakiya
5. jjala
6. jalada
7. brahmavada
8. devadara
9. chraGavidy

Of these, only the Zaunakiya (AVS) and the Paippalda (AVP) recensions have survived. The core Paippalda text is considered earlier than the Zaunakiya, but both also contain later additions and corruptions. In places where the Zaunakiya and the Paippalda agree, it is likely the original version. Often, the two recensions in corresponding hymns have a different verse order, or either has additional verses missing from the other.

Additionally, from the Visnu and Vyu Puranas (older Hindu texts on the gods, goddesses and their histories) it may be possible to glean a few more ancient schools that were not listed in the *Cara Gavyuha*.

These are:

- sumantu
- kabandha
- kumuda
- aulkyana
- babhravya
- munjakea
- saindhavyana
- nakatrakalpa
- ntikalpa
- sachitvidhi

At least some of these may have evolved into the other schools mentioned in the Cara Gavyuha list. Sachitvidhi, Zntikalpa and Nakatrakalpa are the five *kalpa* texts adduced to the Zaunakiya tradition and not separate schools of their own.

There are two main circum-vedic texts associated with the AV, the vaitna sktra and the kauika sktra. These serve the same purpose as the vidhna of the Rigveda and are of greater value in studying the Puranic-Vedic link than the AV text itself There are several Upanishads that are associated with the AV, but appear to be relatively late additions to the tradition. The most important amongst these are the munaka and the pracna Upanishads. The former contains an important reference to Zaunaka, the founder of the Shaunakiya shakha, the latter one is associated with the Paippalda shakha.

***Issues of Note***

- The AV is the first Indic text dealing with medicine. It identifies the causes of disease as living causative agents such as the yatudhnya, the kimidi, the kimi and the duryama. The atharvns seek to kill them with a variety of drugs in order to counter the disease. This approach to disease is surprisingly advanced compared to the trimanual theory developed in the puranic era. Remnants of the original atharvnic thought did persist in the puranic era as can be seen in sucruta's medical treatise garu.
- Here following the atharvn theory the puranic text suggests germs as a cause for leprosy. In the same chapter sucruta also expands on the role of helminths in disease. These two can be directly traced back to the AV sachita. The hymn AVI. 23-24 describes the disease leprosy and recommends the rajaniaucadhi for its treatment. From the description of the aucadhi as black branching entity with dusky patches, it is very likely that is a lichen with antibiotic properties. Thus the AV can stake the claim for being one of the earliest texts to record uses of the antibiotic agents.
- The AV also informs us about warfare. A variety of devices such as the an arrow with a duct for poison (apskambha) and castor bean poison, poisoned net and hook traps, use of disease spreading bugs and smoke screens find a place in the AV sachita (eg. hymns IX.9, IX. 10, the triacdi and nyrbudi hymns). These references to military practices and associated katriya rites were what gave the AV its formidable reputation. In the Mahabharata era that shortly followed after the end of the atharvn period there is a frequent comparison between weapons and the mantras of the heroes. Probably, this comparison was initially supposed to mean the application of deadly weapons as mentioned in the atharvn tradition.
- Several regular and special rituals of the Aryans are a major concern of the AV, just as in the three other

vedas. The major rituals covered by the AV are marriage in kc.

- There are also hymns that are specific to rituals of the bhigu-agnirasas, vaiyas and katriyas. One of the most important of these rites is the Vishsahi Vrata, performed to invoke the Indra and Visnu with the mantras of the XVIIth kc.
- The Vatya rituals were performed by individuals who took on a nomadic ascetic way of living and were generally sent into neighboring states by the ruler of a particular state. They appear to have served a role in reconnaissance and negotiations with neighboring states (compare with Arjuna's Vatya-like journey into the Yadu principality to woo Subhadra). Finally, there are some rituals aimed at the destruction of the enemies (Abhichrika hymns and rites) particularly using the closing mantras of the XVIth kc.
- While these support traditional negative views on the AV, in content they are mirrored by several other hymns from the Rig as well as the Yajuces. Moreover, Abhichrika rites were an integral part of the vedic as amply attested in the brhmana literature.
- Thus the AV as such began fully within the classic vedic fold, though it was more specific to certain clans of fire priests. The development of the Abhichrika rites to their more 'modern' form was seen only in the vidhna literature and in fact began within the Rigvedic tradition in the form of the aigvidhna. The author of the aigvidhna provides passing reference to the development of similar rites in the AV tradition (the references to the Agnirasa Kritys). These rites reached their culmination in the Kauaika and Vaitana Sktra and in some of the Pariaiahthas (appendices) of the atharvn literature. However, these are far removed from the actual hymns themselves suggesting that they represent an encrustation on the atharvanic practice rather than its original form. While in its most extreme form Atharvanic Abhichrika faded away, it did seed the mainstream

Hindu culture resulting in the origin of the Puranic form of the fire ritual (yagas). It also provided the launching pad for the worship of late evolving popular deities like Kumara and Ganapati to capture the mainstream Hindu ritual.

***Philosophical Excursions:*** The AV made the most important contributions to Aryan philosophical thought of all the Sachits. One of the most spectacular expressions of this is seen in the hymn XIII, the Hymn to goddess Earth or the Prithvi Skktam used in the Aghryana rite. The foundations of Vaiceaik, the highest of the Hindu Darmanas is expressed in the mantra XII. 1. 26 in which the atoms (Pmsu) are described forming the stone, the stones agglutinating to form the rocks and the rocks held together to form the Earth. An early pantheistic thought (somewhat convergent to the latter day Viciathadvaitins) is seen in the hymn X.7 that describes the common thread running through all manifest and unmanifest existence as the skacbha. This skacbha is described as what poured out of the Hiranya Garbha, that was the precursor of the complex world in a very simple form (X.7.28). (*Hiranya Garba)* The radiant or golden egg or womb. Esoterically the luminous 'fire mist' or ethereal stuff from which the Universe was formed.")

This Skambha is Indra and Indra is the Skambha which describes all existence. The hymn also describes a pantheistic nature of the Vedic gods (X.7.38): skacbha is the heat that spreads through the universe (Bhkvana) as waves of water; the units of this spreading entity are the gods even as branches of one tree. This one theme that repeatedly presents itself in various interpretations that abounded in later Hindu philosophies and can be considered one of the most fundamental expression of Vedic thought.

***Dating:*** From alleged internal astronomical references (AVS XI.7), it has been surmised that the Atharvanic period included the time when the Pleiades occupied the spring equinox (roughly 2200 BC). Further, tradition suggests that paippalda, one of the early collators, and Vaidharbh, one of the late contributors associated with the Atharvanic text, lived during the reign of prince Hiranyanabha of the Ikshvku dynasty, interpreted to mean that the core AV composition was at least complete by 1500 BC.

While these approaches are not widely accepted as valid, it is clear that the core text of the AV is not particularly recent in the Vedic Sachit tradition, and falls within the classical *Mantra period* of Vedic Sanskrit in the late 2nd millennium BC-roughly contemporary with the Yajurveda mantras, the Rigvedic Khilani, and the redaction of the Samveda.

The Atharvaveda is also the first Indic text to mention Iron (as ayma ayas, literally "black metal"), so that scholarly consensus dates the bulk of the Atharvaveda hymns to the early Indian Iron Age, corresponding to the 12th to 10th centuries BC or the early Kuru kingdom.

During its oral tradition, however, the text has been corrupted by later additions considerably more than the other Vedas, and it is only from comparative philology of the two surviving recensions that the original reading may hoped to be approximated.

## The Brahamans

The Brahmanas are part of the Hindu Shruti; They are composed in Vedic Sanskrit, and the period of their composition is sometimes referred to as the Brahmanic period or age (approximately between 900 BC and 500 BC). They are essentially commentaries of the Vedas, explaining Vedic ritual. The earliest Brahmanas may have been written several centuries earlier, contemporary to the Black Yajurveda commentary prose, but they have only survived in fragments.

Each Brahmana is associated with one of the four Vedas, and within the tradition of that Veda with a particular shakha or school:

***Rigveda***

- Shakala shakha: Aitareya Brahmana (AB)
- Bashkala shakha: Kaushitaki Brahmana (KS)

***Samaveda***

- Kauthuma: PB, Sadyb
- Jayminiya: Jayminiya Brahmana (JB)

***Yajurveda***

- Krishna: the Brahmanas are integrated into the samhitas:
- Maitrayani (MS)
- Charakakatha (CS)
- Kapisthalakatha (KS)
- Taittiriya (TS). The Taittiriya school has an additional Taittiriya Brahmana (TB)

***Shukla***

- Vajasaneyi Madhyandina: Shatapatha Brahmana, Madhyadina recension (ShB)
- Kanva: Shatapatha Brahmana, Kanva recension (ShBK)

***Atharvaveda***

- Paippalada: Gopatha Brahmana

## Aranyakas

The Aranyakas are part of the Hindu sruti; these religious scriptures are sometimes argued to be part of either the Brahmanas or Upanishads. The name translates to "the forest books", meaning, treatises for hermits or sadhus living in the wilderness. This contrasts with the grhyasutras, treatises intended for domestic life. Their language is early Classical Sanskrit, and together with the bulk of the Upanishads, the Aranyakas form the basis of Vedanta, roughly dating to a few centuries on either side of the beginning of the Common Era.

The Aranyakas discuss philosophy and sacrifice. They are believed to have originated with the various mystical ascetic groups that developed in post-Vedic India. The Aranyakas constitute a more philosophical and mystical interpretation of the themes presented in the Vedas, as opposed to the Brahmanas, which were primarily concerned with the proper performance of ritual. Like the Upanishads, the Aranyakas may have initially constituted a secret or hidden teaching, not in the sense of being forbidden or restricted, but rather being both a non-obvious expansion on the themes of the Vedas and a teaching that was primarily conveyed individually from teacher to student.

The Aranyakas are associated with and named after individual Vedic shakhas.

- Aitareya Aranyaka
- Taittiriya Aranyaka
- Katha Aranyaka
- Kaushitaki Aranyaka

## The Upnishad

The Upanishads are part of the Hindu *Shruti* scriptures which primarily discuss meditation and philosophy and are seen as religious instructions by most forms of Hinduism. The very edifice of Indian religions (Hinduism, Jainism, and Buddhism) is built on the strong foundation of the Upanishads.

The Upanishads are mystic or spiritual contemplations of the Vedas, their putative end and essence, and thus known as *Vednta* ("the end of the Vedas"). The Sanskrit term *upanisad* derives from *upa*-(near), *ni*-(down) and *sad* (to sit), *i.e.* referring to the "sitting down near" a spiritual teacher (guru) in order to receive instruction in the Guru-shishya tradition or parampara. The teachers and students appear in a variety of settings (husband answering questions about immortality, a teenage boy being taught by Yama, etc.). Sometimes the sages are women and at times the instructions (or rather inspiration) are sought by kings. Most disciples are men, but not always. A woman disciple, Gargi Vacaknavi, drove the eminent sage Yajnavalkya to fury with her persistent unanswerable questions. The Upanishads have been dated to around 4th century BC. The oldest upanishads, *i.e.* the brihadaranyaka and chandogya upanishad, have been dated much older around the vedic period.

***The Major Upanishads:*** Different Upanishads are formally affiliated with the four Vedas (Rigveda, Yajurveda, Samaveda and Atharvaveda). The Upanishads were transmitted orally by the schools of Vedic recitation sakhas. The longest and oldest Upanishad are the Bahadranyaka and the Chndogya.

The language of the Upanishads is Sanskrit, the oldest among them still classifying as late Vedic Sanskrit. The oldest Upanishads, the Bahadranyakas and the Chndogya are composed in prose. These early texts may date back to the 8th-7th centuries BCE. Later followed a series of Upanishads composed in verse, such as the Katha, and bvetcvatara Upanishads. Reputedly, there were over two hundred *Upanishads*, but the philosopher and commentator Shankara only composed commentaries to eleven of them. The Upanishads commented on by Shankara are generally regarded as the oldest ones. The Muktika Upanishad lists 108 Upanishads. In 1656, at the order of Dara Shikoh, the Upanishads were translated from Sanskrit into Persian.

These philosophical and meditative tracts form the backbone of Hindu thought. Of the early Upanishads, the Aitareya and Kausitki belong to the Rig Veda, Kena and Chndogya to the Samaveda, and Bwhadrasyaka to the Yajurveda, and Pracna and Mundaka to the Atharvaveda. (Associated Upanishad and Vedic book information taken from Radhakrishnan Indian Philosophy, Vol. 1.) In addition, the Mundukya, Katho, bvetcvatara are very important. Others also include Mahnryana and Maitreyi Upanishads as key.

***Place in the Hindu Canon:*** Scholarly breakdowns of the Vedic books see the four Vedas as poetic liturgy, collectively called mantra or samhit, adoration and supplication to the deities of Vedic religion, in parts already sort of melded with monist and henotheist notions, and an over arching order that transcended even the Gods. The *Brahmana* were a collection of ritual instructions, books detailing the priestly functions (which first were available to all men, and so concretized into strictly Brahmin privilege). These came after the Mantra.

In Vedanta, we have the *Aranyakas* and Upanishads. The Aranyaka ("of the forest") detail meditative yogic practices, contemplations of the mystic one and the manifold manifested principles. The Upanishad basically realized all the monist and universal mystical ideas that started in earlier Vedic hymns, and have exerted an influence unprecedented on the rest of Hindu and Indian philosophy.

However, by adherents they are not considered philosophy alone, and form meditations and practical teachings for those advanced enough to benefit from their wisdom.

The Taittiriya Upanishad says this in the Ninth Chapter:

> He who knows the Bliss of *Brahman*, whence words together with the mind turn away, unable to reach It? He is not afraid of anything whatsoever. He does not distress himself with the thought: "Why did I not do what is good? Why did I do what is evil?". Whosoever knows this regards both these as Atman; indeed he cherishes both these as Atman. Such, indeed, is the Upanishad, the secret knowledge of Brahman.

The Upanishads hold information on basic Hindu beliefs, including belief in a world soul, a universal spirit, Brahman, and an individual soul, Atman (Smith 10). A variety of lesser gods are seen as aspects of this one divine ground, Brahman (different from Brahma). Brahman is the ultimate, both transcendent and immanent, the absolute infinite existence, the sum total of all that ever is, was, or ever shall be.

For Advaita philosophers Brahman is not a God in the monotheistic sense, as they do not ascribe to it any limiting characteristics, not even those of being and non-being, and this is reflected in the fact that in Sanskrit, the word brahman has no gender (masculine or feminine or neuter). Dvaita philosophy holds that Brahman is ultimately a personal God, Vishnu, or Krishna (*brahmano hi pratisthaham*, Bhagavad Gita 14.27).

"Who is the Knower?" "What makes my mind think?" "Does life have a purpose, or is it governed by chance?" "What is the cause of the Cosmos?" The sages of the Upanishad try to solve these mysteries and seek knowledge of a Reality beyond ordinary knowing. They also show a preoccupation with states of consciousness, and observed and analyzed dreams as well as dreamless sleep.

***The Philosophy of the Upanishad:*** Due to their mystical nature and intense philosophical bent that does away with all ritual and completely embraces principals of One Brahman and the inner Atman, the Upanishads have a universal feel that has led to their explication in numerous manners, giving birth to the three schools of Vedanta.

Vedantin philosopher Adi Shankara summed up all the Upanishad in one phrase and said that in the end, the ultimate, formless, inconceivable Brahman is the same as our soul, Atman. We only have to realize it through discrimination and piercing through Maya.

A distinctive quotation that is indicative of the call to self-realization, one that inspired Somerset Maugham in titling a book he wrote on Christopher Isherwood, is as follows:

*Get p! Wake up! Seek the guidance of an*

*Illumined teacher and realize the Self.*

*Sharp like a razor's edge is the path,*

*The sages say, difficult to traverse.*

Death Instructing Nachiketa in the *Katho (Word) Upanishad*

The Upanishads also contain the first and most definitive explications of aum as the divine word, the cosmic vibration that underlies all existence and contains multiple trinities of being and principles subsumed into its One Self. The *Isha* says of the Self (Verses 6, 7 & 8 of Isha Upanishad):

*Whoever sees all beings in the soul*

*and the soul in all beings*

*does not shrink away from this.*

*In whom all beings have become one with the knowing soul*

*what delusion or sorrow is there for the one who sees unity?*

*It has filled all.*

*It is radiant, incorporeal, invulnerable,*

*without tendons, pure, untouched by evil.*

*Wise, intelligent, encompassing, self-existent,*

*it organizes objects throughout eternity.*

"*Aum Shanti Shanti Shanti*" This, too, is found first in the Upanishads, the call for tranquility, for divine stillness, for Peace everlasting.

Dara Shikoh, the Muslim sufi, and son of Mughal emperor Shah Jahan, translated the Upanishads in Persian in order to find in it elements of monotheism that might pave the way for a common mystical language between Islam and Hinduism.

The following is a list of the ten "principal" (mukhya) Upanishads that were commented upon by Shankara, and that are accepted as shruti by all Hindus. They are listed with their associated Veda (Rigveda (ZV), Samaveda (SV), White Yajurveda (ZYV), Black Yajurveda (KYV), Atharvaveda (AV)).

1. Aitareya (ZV)
2. Bahadranyaka (ZYV)
3. ca (ZYV)
4. Taittiraya (KYV)
5. Kamha (KYV)
6. Chndogya (SV)
7. Kena (SV)
8. Mug
9. aka (AV)
10. MG
11. kkya (AV)
12. Prayna (AV)

The Kauaitki, Zvetavatara and Maitryani Upanishads are sometimes added to extend the canon to 12 or 13. They are also the oldest Upanishads, likely all of them dating to before the Common Era. From linguistic evidence, the oldest among them are likely the Bwhadranyaka and Chndogya Upanishads, belonging to the late Vedic Sanskrit period; the remaining ones are at the transition from Vedic to Classical Sanskrit.

***Canon by Vedic Shakha:*** The older Upanishads are associated with Vedic Charanas (Shakhas or schools). The Aitareya Upanishad with the Shakala shakha, the Kauaitki Upanishad with the Bashakala shakha; the Chndogya

Upanishad with the Kauthuma shakha, the Kena Upanishad, and the Jaiminiya Upanishad Brahmana, with the Jaiminiya shakha; the Kamha Upanishad with the Caraka-Katha shakha, the Taittiraya and Zvetavatara with the Taittiriya shakha; the Maitryani Upanishad with the Maitrayani shakha; the Bwhadranyaka and *ca Upanishads with the Vajasaneyi Madhyandina shakha, and the MG.

*Kkya and Mug:* aka Upanishads with the Shaunaka shakha. Additionally, parts of earlier texts, of Brahmanas or passages of the Vedas themselves, are sometimes considered Upanishads.

***The Muktika Canon:*** The following is a list of the 108 canonical Upanishads of the Advaita school, according to the Muktika Upanishad (number 108), 1:30-39 (which does not list the associated Veda). In this canon,

- 10 Upanicads are associated with the Rigveda and have the Shnti beginning *vanme-manasi*.
- 16 Upanicads are associated with the Samaveda and have the Shnti beginning *pyyantu*.
- 19 Upanicads are associated with the White Yajurveda and have the Shnti beginning *pkrnamada*.
- 32 Upanicads are associated with the Black Yajurveda and have the Shnti beginning *sahanvavatu*.
- 31 Upanicads are associated with the Atharvaveda and have the Shnti beginning *bhadram-karnebhis*.

The first 10 are grouped as *mukhya* "principal", and are identical to those listed above. 21 are grouped as Smnya Vednta "common Vedanta", 23 as Sannysa, 9 as Shkta, 13 as Vaishnava, 14 as Shaiva and 17 as Yoga Upanishads.

1. "The Inner Ruler"
2. "Who moves the world?"
3. Death as Teacher"
4. "The Breath of Life"
5. Mug
6. "Two modes of Knowing"

7, 8. MG

9. "Consciousness and its phases"
10. "From Food to Joy"
11. "The Microcosm of Man"
12. "Song and Sacrifice"
13. Bahadranyaka (ZYV, Mukhya)
14. Brahma (KYV, Sannyasa)
15. Kaivalya (KYV, Shaiva)
16. Jbla (ZYV, Sannyasa)
17. "The Faces of God"
18. Hassa (ZYV, Yoga)
19. runeya (SV, Sannyasa)
20. Garbha (KYV, Sannyasa)
21. Nryana (KYV, Vaishnava)
22. Paramahacsa (ZYV, Sannyasa)
23. Amatabindu (KYV, Yoga)
24. Amatanda (KYV, Yoga)
25. Zira (AV, Shaiva)
26. Atharvaikha (AV, Shaiva)
27. Maitryani (SV, Sannyasa)
28. Kauaitki (ZV, Samanya)
29. Bwhajjbla (AV, Shaiva)
30. Nasishatpan (AV, Vaishnava)
31. Klgnirudra (KYV, Shaiva)
32. Maitreyi (SV, Sannyasa)
33. Subla (ZYV, Samanya)
34. Kcurika (KYV, Yoga)
35. Mantrika (ZYV, Samanya)
36. Sarvasra (KYV, Samanya)
37. Nirlamba (ZYV, Samanya)
38. Zukarahasya (KYV, Samanya)
39. Vajraskchi (SV, Samanya)
40. Tejobindu (KYV, Sannyasa)
41. Ndabindu (ZV, Yoga)

42. Dhynabindu (KYV, Yoga)
43. Brahmavidy (KYV, Yoga)
44. Yogatattva (KYV, Yoga)
45. Tmabodha (ZV, Samanya)
46. Parivrt (Nradaparivrjaka) (AV, Sannyasa)
47. Triaikhi (ZYV, Yoga)
48. Sit (AV, Shakta)
49. Yogachk
50. Mani (SV, Yoga)
51. Nirvna (ZV, Sannyasa)
52. Mag
53. Alabrhmana (ZYV, Yoga)
54. Dakcinmkrti (KYV, Shaiva)
55. Zarabha (AV, Shaiva)
56. Skanda (Trip
57. Vibhkmi) (KYV, Samanya)
58. Mahnryana (AV, Vaishnava)
59. Advayatraka (ZYV, Sannyasa)
60. Rmarahasya (AV, Vaishnava)
61. Rmatpani (AV, Vaishnava)
62. Vsudeva (SV, Vaishnava)
63. Mudgala (ZV, Samanya)
64. ZG
65. Ilya (AV, Yoga)
66. Paiegala (ZYV, Samanya)
67. Bhikcu (ZYV, Sannyasa)
68. Mahad (SV, Samanya)
69. Zriraka (KYV, Samanya)
70. Yogaaikh (KYV Yoga)
71. Turiytita (ZYV, Sannyasa)
72. Sannysa (SV, Sannyasa)
73. Paramaha Csaparivrjaka (AV, Sannyasa)
74. Akcamlika (Mlika) (ZV, Shaiva)

75. Avyakta (SV, Vaishnava)
76. Ekkcara (KYV, Samanya)
77. Annapkrna (AV, Shakta)
78. Skrya (AV, Samanya)
79. Akci (KYV, Samanya)
80. Adhytm (ZYV, Samanya)
81. Kug
82. Ika (SV, Sannyasa)
83. Svitri (SV, Samanya)
84. Tm (AV, Samanya)
85. Paupata (AV, Yoga)
86. Parabrahma (AV, Sannyasa)
87. Avadhkta (KYV, Sannyasa)
88. Devi (AV, Shakta)
89. Tripurtapani (AV, Shakta)
90. Tripura (ZV, Shakta)
91. Kamharudra (KYV, Sannyasa)
92. Bhvana (AV, Shakta)
93. Rudrahadaya (KYV, Shaiva)
94. Yogakun
95. Alini (KYV, Yoga)
96. Bhasma (AV, Shaiva)
97. Rudrkca (SV, Shaiva)
98. Ganapati (AV, Shaiva)
99. Daraana (SV, Yoga)
100. Trasra (ZYV, Vaishnava)
101. Mahvkya (AV, Yoga)
102. Pancabrahma (KYV, Shaiva)
103. Prngnihotra (KYV, Samanya)
104. Goplatpani (AV, Vaishnava)
105. Kacga (AV, Vaishnava)
106. Yjnavalkya (ZYV, Sannyasa)
107. Varha (KYV, Sannyasa)

108. Zmyyani (ZYV, Sannyasa)
109. Hayagriva (AV, Vaishnava)
110. Datttreya (AV, Vaishnava)
111. Gru
112. A (AV, Vaishnava)
113. Kali-Sagmrana (Kali) (KYV, Vaishnava)
114. Jbla (SV, Shaiva)
115. Saubhgya (ZV, Shakta)
116. Sarasvatirahasya (KYV, Shakta)
117. Bahvaca (ZV, Shakta)
118. Muktika (ZYV, Samanya)

## Smirti

Smriti refers to a specific body of Hindu religious scripture. *Smriti* also denotes non-*Shruti* texts generally, seen as secondary in authority to *Shruti*.

***Categorization:*** Classifying the Smriti has been a contentious issue, even the names of proposed categories are debated. One such taxonomy follows:

1. Dharmasastra or the laws. This is represented by 18 books. Each book corresponds to an age of time.
2. Itihasa or the histories. This is represented by 4 books. It includes the Mahakavyas, or Epics Mahabharata and the Ramayana.
3. Purana or the writings. This is represented by 18 books. They are secondary scriptures that mainly focus on Vishnu or Shiva as the preferred supreme Deity.
4. Sutras or proverbs. This is represented by 6 categories of documents: the Shiksha, Vyakarana, Chandas, Nirukta, Jyotisha, & Kalpa.
5. Agama or the doctrines. There are three major divisions by doctrine: the Vaisnava, Saiva, and Sakta. Another way of grouping them is by Mantra, Tantra, and Yantra.
6. Darsana (aka Dyasana) or philosophies. This is represented by 6 schools of thought: Nyaya, Vaisheshika,

Sankhya, Yoga, Purva Mimamsa, & Uttara Mimamsa (aka Vedanta).

***Itihaas:*** Itihasa is the word for *History*. The word is also used in most other Indian languages, with the same meaning. The word is often used within India to refer to Hindu Epics, (aka Mahakavyas) such as the Ramayana, the Mahabharata. It includes the Yogavasishtha and Harivamsa.

***Itihasa can also Refer to the Puranas:*** Some historians may also accept this use, with a view to attribute historic veracity to such legends. This usage is eschewed by most historians because the dating of events alluded to in these legends, as also the substantiation of their historic veracity and accuracy, is generally impossible.

***Chronology:*** The Itihasa-Purana is mentioned in the Chandogya Upanishad. There are also references to the Puranas in the Atharvaveda 11.7.24 and the Satapatha Brahmana 11.5.6.8. and 13.4.3.13. There are probably also references to the Itihasa in the Nirukta. The Itihasa-Purana was considered to be almost as sacred and ancient as the Vedas. However, the ancient Itihasa-Purana was probably different from the later Itihasa and Puranas. Scholars argue that some contents of the Puranas may date to an earlier period.

## Mahabharata

The is one of the two major Sanskrit epics of ancient India, the other being the *Ramayana*. With more than 74,000 verses, plus long prose passages, or some 1.8 million words in total, it is one of the longest epic poems worldwide. It is also of immense religious and philosophical importance in India, in particular for including the Bhagavad Gita, an important text of Hinduism. The title may be translated as "the great tale of the Bharata Dynasty", according to the Mahabharata's own testimony extended from a shorter version simply called *Bhrata* of 24,000 verses The epic is part of the Hindu *itihsas*, literally "that which happened", along with the *Ramayana* and the *Purnas*.

Traditionally, the Mahabharata is ascribed to Vyasa. Due to its immense length, its philological study has a long history of attempting to unravel its historical growth and composition layers. In its final form, it is assumed to have been completed between the 3rd and 5th centuries, with its central core (consisting of only a fraction of the full 1.8 million words) going back as far as the 5th century BC.

***Influence:*** With its vast philosophical depth and sheer magnitude, a consummate embodiment of the ethos of not only India but of Hinduism and Vedic tradition, the Mahabharata's scope and grandeur is best summarized by one quotation from the beginning of its first *parva* (section): "What is found here, may be found elsewhere. What is not found here, will not be found elsewhere." In its scope, the Mahabharata is more than simply a story of kings and princes, sages and wisemen, demons and gods; its author, Vyasa, says that on of its aims is elucidating the four goals of life: kama (pleasure), artha (wealth), dharma (duty) and moksha (liberation). The story culminates in moksha, believed by many Hindus to be the ultimate goal of human beings. Karma and dharma play an integral role in the Mahabharata.

The Mahabharata includes large amounts of Hindu mythology, cosmological stories of the gods and goddesses, and philosophical parables aimed at students of Hindu philosophy. Among the principal works and stories that are a part of the Mahabharata are the following (often considered isolated as works in their own right):

- Bhagavad Gita (Krishna instructs and teaches Arjuna. Anusasanaparva.)
- Damayanti (or Nala and Damayanti, a love story. Aranyakaparva.)
- Krishnavatara (the story of Krishna, the *Krishna Leela*, which is woven through many chapters of the story)
- Rama (an abbreviated version of the Ramayana. Aranyakaparva.)
- Rishyasringa (also written as Rshyashrnga, the horned boy and rishi. Aranyakaparva.)
- Vishnu sahasranama (the most famous hymn to Vishnu, which describes His 1000 names; Anushasanaparva.)

***Textual History and Organization:*** It is undisputed that the full length of the Mahabharata has accreted over a long period. The Mahabharata itself (1.1.61) distinguishes a core portion of 24,000, the *Bharata* proper, as opposed to additional "secondary" material, and the Ashvalayana Grhyasutra (3.4.4) makes a similar distinction. Not unlike the field of Homeric studies, research on the Mahabharata has put an enormous effort into recognizing and dating various layers within the text. Oldenberg (1922) stipulated that the supposed original poem once carried an immense "tragic force", but dismissed the full text as a "horrible chaos".

The first testimony of the existence of the full text is an inscription at Khoh, dated to 533 BC, describing the Mahabharata as a "collection of 100,000 verses" (*shatasahasri samhita*). The redaction of this large body of text was carried out after formal principles, emphasizing the numbers 18 and 12. The addition of the latest parts may be dated by the absence of the *Anushasana-parva* from *MS Spitzer*, the oldest surviving Mahabharata manuscript dated to ca. 200 BC. From this evidence, it is likely that the redaction into 18 books took place in the 3rd or 4th century BC. An alternative division into 20 parvans appears to have coexisted for some time. The division into 100 sub-parvans (mentioned in Mbh. 1.2.70) is older, and most parvans are named after one of their constituent sub-parvans. The Harivamsha consists of the final two of the 100 sub-parvans, and was considered an appendix (*khila*) to the Mahabharata proper by the redactors of the 18parvans.

The Adi-parvan is dedicated to the snake sacrifice (*sarpasattra*) of Jayamejaya, explaining its motivation, detailing why all snakes in existence were intended to be destroyed, and why in spite of this, there are still snakes in existence. This sarpasattra material was often considered an independent tale added to a version of the Mahabharata by "thematic attraction" (Minkowski 1991), and considered to have particularly close connection to Vedic (Brahmana literature), in particular the Panchavimsha Brahmana which describes the Sarpasattra as originally performed by snakes, among which are snakes named Dhrtarashtra and Janamejaya, two main characters of the Mahabharata's sarpasattra, and Takshaka, the name of a snake also in the Mahabharata.

The 18 division into parvans is as follows:

| Parvan | Title | Sub-parvans | Contents |
|---|---|---|---|
| 1 | *Adi-parva* | 1-19 | Introduction, birth and upbringing of the princes. |
| 2 | *Sabha-parva* | 20-28 | Life at the court, the game of dice, and the exile of the Pandavas. Maya Danava erects the palace and court (*sabha*), at Indraprastha. |
| 3 | *Aranyaka-parva (also Vanaparva, Aranyaparva)* | 29-44 | The twelve years in exile in the forest (*aranya*). |
| 4 | *Virata-parva* | 45-48 | The year in exile spent at the court of Virata. |
| 5 | *Udyoga-parva* | 49-59 | Preparations for war. |
| 6 | *Bhishma-parva* | 60-64 | The first part of the great battle, with Bhishma as commander for the Kauravas. |
| 7 | *Drona-parva* | 65-72 | The battle continues, with Drona as commander. |
| 8 | *Karna-parva* | 73 | The battle again, with Karna as commander. |
| 9 | *Shalya-parva* | 74-77 | The last part of the battle, with Shalya as commander. |
| 10 | *Sauptika-...parva* | 78-80 | How Ashvattama and the remaining Kauravas killed the Pandava army in their sleep (Sauptika). |
| 11 | *Stri.....parva* | 81-85 | Gandhari and the other women (*stri*) lament the dead. |
| 12 | *Shanti-parva* | 86-88 | The crowning of Yudhisthira, and his instructions from Bhishma |

| Parvan | Title | Sub-parvans | Contents |
|---|---|---|---|
| 13 | *Anusasana-parva* | 89-90 | The final instructions (*anusasana*) from Bhishma. |
| 14 | *Ashvamedhika-parva* | 91-92 | The royal ceremony of the ashvamedha conducted by Yudhisthira. |
| 15 | *Ashramavasika-parva* | 93-95 | Dhritarashtra, Gandhari and Kunti leave for an ashram, and eventual death in the forest. |
| 16 | *Mausala-parva* | 96 | The infighting between the Yadavas with maces (*mausala*). |
| 17 | *Mahaprasthanika-parva* | 97 | The first part of the path to death (*mahaprasthana* "great journey") of Yudhisthira and his brothers. |
| 18 | *Svargarohana-parva* | 98 | The Pandavas return to the spiritual world (svarga). |
| *khila* | Harivamsaparva | 99-100 | life of Krishna. |

The Shatapatha Brahmana gives an account of an Ashvamedha performed by Janamejaya Parikshita.

According to Mbh. 1.1.50, there were three versions of the epic, beginning with *Manu* (1.1.27), *Astika* (1.3, sub-parvan 5) or *Vasu* (1.57), respectively. These versions probably correspond to the addition of one and then another 'frame' settings of dialogues. The *Vasu* version corresponds to the oldest, without frame settings, beginning with the account of the birth of Vyasa. The Astika version adds the Sarpasattra and Ashvamedha material from Brahmanical literature, and introduces the name *Mahabharata* and identifies Vyasa as the work's author. The redactors of these additions were probably Pancharatrin scholars who according to Oberlies (1998) likely retained control over the text until its final redaction in the 3rd or 4th century BC. Mention of the Hunas in the *Bhishma-parva* appears to imply that the compilation of the text was still ongoing in 400 BC.

***Historicity:*** The historicity of the events of the story is unclear. Many historians believe it to be a work of fiction. The epic's setting certainly has a historical precedent in Vedic India, where the Kuru kingdom was the center of political power in the late 2nd and early 1st millennia BC. Ancient Indian scholars have calculated chronologies for the Mahabharata war, comparable to the Hellenistic attempts at a chronology of Greek mythology, the 5th century mathematician Aryabhatta arriving at an approximate date for the Kurukshetra battle of 3100 BC. Contentious and disputable attempts to date the events of the Mahabharata with the help of archaeoastronomy have claimed dates in the 6th millennium BC. According to the Puranas, there is a time gap of 1015 or 1500 years between Parikshit's birth during the Mahabharata war and the coronation of king Mahapadma Nanda (ca. 364-382 BC).

***Synopsis:*** The epic employs the 'tale-within-tale' structure popular in many Indian religious and secular works. It is recited to the King Janamejaya by Vaishampayana, a disciple of Vyasa. The core story of the work is that of a dynastic struggle for the throne of Hastinapura, the kingdom ruled by the Kuru clan. The two collateral branches of the family that participate in the struggle are the Kauravas, the elder branch of the family, and the Pandavas, the younger branch.

The struggle culminates leading to the Great battle of Kurukshetra, and the Pandavas are ultimately victorious. The Mahabharata itself ends with the death of Krishna, and the subsequent end of his dynasty, and ascent of the Pandava brothers to Heaven. It also marks the beginning of the Hindu age of Kali (Kali Yuga), the fourth and final age of mankind, where the great values and noble ideas have crumbled, and man is speedily heading toward the complete dissolution of right action, morality and virtue. Some of the most noble and revered figures in the Mahabharat end up fighting on the side of the Kauravas, due to conflicts of their dharma, or duty. For example, Bhishma had vowed to always protect the king of Hastinapur, whoever he may be. Thus, he was required to fight on the side of evil knowing that his Pandavas would end up victorious only with his death.

***Authorship:*** The epic is traditionally ascribed to Maha Rishi Veda Vyasa, who is one of the major dynastic characters within the epic. The first section of the Mahabharata states that it was Ganesha who, at the behest of Vyasa, fixed the text in manuscript form. Lord Ganesha is said to have agreed, but only on condition that Vyasa never pause in his recitation. Vyasa then put a counter-condition that Ganesha understand whatever he recited, before writing it down. In this way Vyasa could get some respite from continuously speaking by saying a verse which was difficult to understand. This situation also serves as a popular variation on the stories of how Ganesha's right tusk was broken (a traditional part of Ganesha imagery). This version attributes it to the fact that, in the rush of writing, the great elephant-headed divinity's pen failed, and he snapped off his tusk as a replacement in order that the transcription not be interrupted.

***Bhishma:*** Janamejaya's ancestor Shantanu, the king of Hastinapur has a short-lived marriage with the goddess Ganga and has a heroic son, Bhishma. Many years later, when the king goes hunting, he spots Satyavati, the daughter of a fisherman and wants to marry her. The fisherman says that he will agree only on one condition-that her son should ascend the throne of Hastinapur. Bhishma, his son by his first wife, renounces his right to the throne and vows to be a lifelong celibate, so that Satyavati's son can be the king. Such a vow was unheard of amongst warrior dynasties thus inspiring the name Bhishma-'the person of the terrible oath'.

***The Pandavas:*** Unfortunately Satyavati's sons die young and her grandson Pandu ascends the throne as his elder brother Dhritarashtra is blind. But this king is cursed with a terrible fate-if he ever makes love to a woman, he will die. He retires to the forest along with his two wives. Using a magical spell to summon the gods, his elder queen Kunti gives birth to three sons Yudhishthira, Arjuna and Bhima. His younger queen, Madri bears the twins Nakula and Sahadeva. Pandu and Madri cannot resist temptation, and die in the forest and Kunti returns to Hastinapura with her sons. The rivalry between the Pandavas and the Kauravas start from childhood itself. Dhritarashtra's sons, the Kauravas, led by the eldest Duryodhana, detest their cousins. However, they were the favorite of their teacher Drona and (the Pandavas) grow up to be exceptional. Each one of the Pandavas is said to have one exceptional strength or virtue-Yudhishthira is the wisest and most virtuous, Arjuna the bravest warrior, Bhima the strongest and the twins Nakula and Sahadeva are endowed with exceptional beauty.

When the princes come of age, a tournament is held to display the strength and specialities of the princes of Hastinapur. When Arjuna was hailed as a master of archery, a young man challenges him for a duel. He declares his name is Karna, and he was the son of the charioteer. When asked to prove that his birth is of royal birth, which is the criterion for joining the tournament, Duryodhana, spotting a potential ally, jumps over to his side and gives his kingdom of Angawardana. Karna is forever grateful for this act. Because of this, he becomes Duryodhana's closest friend and plays a crucial role in the upcoming war.

***Draupadi:*** Meanwhile Duryodhana plots to get rid of the Pandavas and tries to kill the Pandavas secretly by burning their palace which is made of lac. However, the Pandavas are warned and escape from the palace. They live in hiding for sometime. In course of this exile Arjuna wins the hand of the Panchala princess Draupadi. When he returns with his bride, Arjuna goes to his mother to show her his prize, exclaiming, "Mother, see what I have won!". Kunti, not noticing the princess, tells Arjuna that whatever he has won must be shared with his brothers. To ensure that their mother never utters a falsehood even by mistake, the brothers take her as a common wife. All of the Pandavas love Draupadi dearly. In some interpretations, Draupadi alternates months or years with each brother. At this juncture they also meet Krishna, a close friend of Draupadi, who would become their lifelong ally and guide.

***Duryodhana's Game:*** Duryodhana, who now has a friend in the peerless warrior Karna, resigns to the coming back of the Pandavas with their new royal ally. Soon they conquer the whole of India and its adjoining regions and Yudhishthira is crowned the emperor. This proves too much for Duryodhana who feels death would be better than watching one's foes prosper. His maternal uncle Shakuni, convinced that however brave his nephew may be, he was no match for his cousins, decides to use a ruse to destroy the Pandavas. He forces Dhritarashtra to invite the Pandavas for a game of dice in which he wins everything from Yudhishthira, including himself, his brothers and Draupadi through the use of a trick.

The jubilant Kauravas insult them in their helpless state and even try to strip Draupadi. Her honour is saved by the grace of Krishna. When the elders intervene and Dhritarashtra has to restore everything to the Pandavas, Shakuni forces another game of dice which he again wins. The Pandavas are required to go into exile for 13 years, and on the 13th year they must remain hidden. If discovered by the Kauravas, they will be forced into exile for another 13 years.

***The Battle at Kurukshetra:*** When the Pandavas after many hardships and exile request for at least five villages for the five brothers from their vast kingdom, Duryodhana refuses to give in. Krishna goes to broker peace but fails. War was inevitable. The two sides summon vast armies to their help and line up at Kurukshetra for a war. The Kingdoms of Dwaraka, Kasi, Kekaya, Magadha, Matsya, Chedi, Pandya and the Yadus of Mathura and some other clans like the Parama Kambojas from Transoxiana were allied with the Pandavas; the allies of the Kauravas comprised the kings of Pragjyotisha, Anga, Kekaya (Kekaya brothers who were enemies of the Kekeya brothers on the Pandava side), Sindhudesa (including Sindhus, Sauviras and Sivis), Mahishmati, Avanti in Madhyadesa, Madras, Gandharas, Bahlikas, Kambojas (with Yavanas, Sakas, Tusharas etc.) and many others. Seeing himself facing grandsire Bhishma and his teacher Drona on Duryodhana's side due to their vow to serve the state of Hastinapur, Arjuna is heartbroken at the idea of killing them.

Krishna who has chosen to drive Arjuna's chariot wakes him up to his call of duty in the famous Bhagavad Gita section of the epic. Though initially sticking to chivalrous notions of warfare, the Kauravas and Pandavas soon descended into dishonourable warfare. At the end of the 18 days slaughter only the Pandavas and Krishna survive with a few old warriors from the Kaurava side.

***The End of the Pandavas:*** Beholding the carnage, the noble mother of the Kauravas, Gandhari who had lost all her sons, curses Krishna to be a witness to a similar annihilation of his family, for though divine and capable of stopping the war, he had not done so. Krishna who had incarnated precisely to destroy the wicked kings accepts the curse, which bears fruit 36 years later. He then departs from the world and the Pandavas who had ruled righteously all along, now tired, decide to renounce everything. Clad in skins and rags they retire to the Himalayas and ascend the peaks. One by one they get exhausted, die and ascend to heaven. Only the virtuous Yudhisthira who had tried everything to prevent the carnage and never very happy with his kingship afterwards, ascends to the heaven in his physical body.

Arjuna's grandson Parikshita rules after them and dies bitten by a snake. His furious son, Janamejaya, decides to perform a snake sacrifice (sarpasattra) in order to destroy the snakes. It is at this sacrifice that the tale of his ancestors is narrated to him.

***Bhagbatgita:*** The Bhagavad Gita is an ancient Sanskrit text comprising of some 700 verses from the epic work known as Mahabharata (*Bhishma Parva* chapters 23 – 40). The verses, using the range and style of Sanskrit meter (poetry) (*chhanda*) with similes and metaphors, are very poetic; hence the title, which translates to "the Song of the Divine One", of Bhagavan in the form of Krishna. It is revered as sacred by the majority of Hindu traditions, and especially so by followers of Krishna. In general speech it is commonly reerred to as *The Gita*

The content of the Bhagavad Gita is a conversation between Krishna and Arjuna taking place on the battlefield of Kurukshetra just prior to the start of a climactic war. Responding to Arjuna's confusion and moral dilemma, Krishna explains to Arjuna his duties and elaborates on number of different Yogic and Vedantic philosophies, with examples and analogies. This has led to the Gita often being described as a concise guide to Hindu philosophy. During the discourse, Krishna reveals his identity as the Supreme Being Himself (Bhagavan), blessing Arjuna with an awe-inspiring glimpse of His divine absolute form. The Bhagavad Gita is also called *Gitopanicad*, implying it to be an 'Upanishad'. While technically it is considered a *Smati* text, it has singularly achieved the status of *sruti*, or *revealed knowledge*.

***Background:*** *The Bhagavad Gita* describes the mind as turbulent and obstinate. 'The Chariot of the Body': The five horses represent the five senses (tongue, eyes, nose, ears and skin). The rein symbolises the mind, the driver is the intelligence, and the passenger is the spirit soul.The discourse on the Bhagavad Gita begins before the start of the climactic battle at Kurukshetra. It begins with the pandava prince Arjuna, as he becomes filled with doubt on the battlefield. Realising that his enemies are his own relatives, beloved friends, and revered teachers, he turns to his charioteer and guide, Sri Krishna (an avatar of Sri Vishnu), for advice.

Krishna counsels Arjuna, beginning with the tenet that the soul is both eternal and immortal. Any 'death' on the battlefield would involve only the shedding of the body but the inner soul is permanent. Krishna goes on to expound on the yogic paths of devotion, action, meditation and knowledge. Fundamentally, the Bhagavad Gita proposes that true enlightenment comes from growing beyond identification with the Ego, the 'False Self', and that one must identify with the Truth of the immortal Self, (the soul or Atman). Through detachment from the material sense of Ego, the Yogi, or follower of a particular path of Yoga, is able to transcend his illusory mortality and attachment from the material world and enter into the realm of the Supreme.

To demonstrate his divine nature, Krishna grants Arjuna the boon of cosmic vision (albeit temporary) and allows the prince to see His 'Universal Form'. He reveals that He is fundamentally both the ultimate essence of *Being* in the universe, and also its material body. This is called the Vishvarupa/Viratrupa.

The Gita refers to the war as Dharma Yuddha, meaning a just war. Chapter 4, verse 7, clearly states that God takes incarnation to establish righteousness in the world.

***Dating of the Text:*** Though it is not exactly clear when the Bhagavad Gita was written, the majority of historians give dates around 50 BCE-500 BCE. There is, however, considerable debate on the subject. Based on the differences in the poetic styles and the external influences such as Patanjali's *Yoga Sutra*, certain scholars have expressed the opinion that Bhagavad Gita was added to the *Mahabharata* at a later date. Again, there are other theories based on astronomical calculations cited in the Mahabharata that place the incidents upon which the Gita is based around the time 5561 BCE. The latter date corresponds with traditional knowledge and reflects the beliefs of many devotional Hindus.

***The Scripture of Yoga:*** The Gita addresses the discord between the senses and the intuition of cosmic order. It speaks of the Yoga of equanimity, a detached outlook. The term Yoga covers a wide range of meanings, but in the context of the Bhagavad Gita, describes a unified outlook, serenity of mind, skill in action, and the ability to stay attuned to the glory of the Self (Atman), which is of the same essence as the basis of Being (Brahman). According to Krishna, the root of all suffering and discord is the agitation of the mind caused by selfish desire. The only way to douse the flame of desire is by simultaneously stilling the mind through self discipline and engaging oneself in a higher form of activity.

However, abstinence from action is regarded as being just as detrimental as extreme indulgence. According to the Bhagavad Gita, the goal of life is to free the mind and intellect from their complexities, and to focus them on the glory of the Self, by dedicating one's actions to the divine. This goal can be achieved through the Yogas of meditation, action, devotion and knowledge. The Gita describes the best Yogi as one who constantly comtemplates God. Krishna summarizes the Yogas through eighteen chapters. There are four kinds of Yoga-Raja Yoga or *Psycho-Physical Meditation*, Bhakti Yoga or *Devotion*, Karma Yoga or *Selfless Action*, and Jnana Yoga or *Self Transcending Knowledge*.

While each path differs, their fundamental goal is the same-to realize Brahman (the ***Divine Essence***) as being the ultimate truth upon which our material universe rests, that the body is temporal, and that the *Supreme Soul* (Paramatman) is infinite. Yoga's aim (nirvana or moksha) is to escape from the cycle of reincarnation through realization of oneness with the ultimate reality. There are three stages to self-realisation enunciated from the Bhagavad Gita:

1. Brahman-The impersonal universal energy
2. Paramatma-The Supreme Soul sitting in the heart of every living entity.
3. Bhagavan-God as a personality, with a transcendental form.

The below are quotations from Krishna concerning the four principle yogas of the Bhagavad Gita:

***On the Goal of Yoga:*** "And whoever, at the time of death, quits his body, remembering Me (Krishna) alone, at once attains My (Krishna's-Transcendental) nature. Of this there is no doubt."

***On Bhakti Yoga:*** Put simply, Bhakti Yoga is *Service in Love and Devotion to God* (Krishna in the context of the Bhagavad Gita).

> *"I consider the Yogi-devotee-who lovingly contemplates on Me with supreme faith, and whose mind is ever absorbed in Me-to be the best of all the Yogis". "After attaining Me, the great souls do not incur rebirth in this miserable transitory world, because they have attained the highest perfection." "... those who, renouncing all actions in Me, and regarding Me as the Supreme, worship Me... For those whose thoughts have entered into Me, I am soon the deliverer from the ocean of death and transmigration, Arjuna. Keep your mind on Me alone, your intellect on Me. Thus you shall dwell in Me hereafter." "And he who serves Me with the yoga of unswerving devotion, transcending these qualities [binary opposites, like good and evil, pain and pleasure] is ready for liberation in Brahman." "Fix your mind on Me, be devoted to Me, offer service to Me, bow down to Me, and you shall certainly reach Me. I promise you*

*because you are My very dear friend." "Setting aside all meritorious deeds (Dharma), just surrender completely to My will (with firm faith and loving contemplation). I shall liberate you from all sins. Do not fear."*

***On Karma Yoga:*** Karma Yoga is essentially *Acting, or doing one's duties in life as per his / her* dharma, *or duty, without concern of results*-a sort of constant sacrifice of action to the Supreme. It is action done without thought of gain. In a more modern interpretation, it can be viewed as duty bound deeds done without letting the nature of the result affecting ones actions. It is said that the results can be of 3 types-as aimed for, opposite of what is aimed for, or a mixture of these. If one can perform his duties (as prescribed in the Vedas) without any anticipation of the result of his actions, he is bound to succeed. It includes, but is not limited to, dedication of one's chosen profession and its perfection to God. It is also visible in community and social service, since they are inherently done without thought of personal gain.

Krishna advocates Nishkam Karma Yoga (the Yoga of Selfless Action) as the ideal path to realize the Truth. Allocated work done without expectations, motives, or thinking about its outcomes tends to purify one's mind and gradually makes an individual fit to see the value of reason and the benefits of renouncing the work itself. These concepts are vividly described in the following verses:

*"To action alone hast thou a right and never at all to its fruits; let not the fruits of action be thy motive; neither let there be in thee any attachment to inaction"*

— verse 47, Chapter 2-Samkhya theory and Yoga practise, The Bhagavadgita-*Radhakrishnan*

*"Fixed in yoga, do thy work, O Winner of wealth(Arjuna), abandoning attachment, with an even mind in success and failure, for evenness of mind is called yoga"*

— verse 48, Chapter 2-Samkhya theory and Yoga practise, The Bhagavadgita-*Radhakrishnan*

*"With the body, with the mind, with the intellect, even merely with the senses, the Yogis perform action toward self-purification, having abandoned attachment. He who*

*is disciplined in Yoga, having abandoned the fruit of action, attains steady peace..."*

In order to achieve true liberation, it is important to control all mental desires and tendencies to enjoy and sense pleasures. The following verses illustrate this:

"When a man dwells in his mind on the object of sense, attachment to them is produced. From attachment springs desire and from desire comes anger."

"From anger arises bewilderment, from bewilderment loss of memory; and from loss of memory, the destruction of intelligence and from the destruction of intelligence he perishes"

***On Jnana Yoga:*** Jnana Yoga is a process of learning to discriminate between what is real and what is not, what is eternal and what is not. Through a steady advancement in realization of the distinction between Real and the Unreal, the Eternal and the Temporal, one develops into a *Jnana Yogi*. This is essentially a path to God through knowledge and discrimination. "When he perceives the various states of being as resting in the One, and from That alone spreading out, then he attains Brahman. / They who know, through the eye of knowledge, the distinction between the field and the knower of the field, as well as the liberation of beings from material nature, go to the Supreme."

***On Raja Yoga:*** Raja Yoga is the stilling of the mind and body through meditating techniques, geared at realizing one's true nature. This practice was later described by Patanjali in his Yoga Sutras.

"To practice yoga, one should go to a secluded place and should lay kusa grass on the ground and then cover it with a deerskin and a soft cloth. The seat should be neither too high nor too low and should be situated in a sacred place. The yogi should then sit on it very firmly and practice yoga to purify the heart by controlling his mind, senses and activities and fixing the mind on one point. One should hold one's body, neck and head erect in a straight line and stare steadily at the tip of the nose. Thus, with an unagitated, subdued mind, devoid of fear, completely free from sex life, one should meditate upon

Me within the heart and make Me the ultimate goal of life. Thus practicing constant control of the body, mind and activities, the mystic transcendentalist, his mind regulated, attains to the kingdom of God [or the abode of Krishna] by cessation of material existence."

***Influence of the Bhagavadgita:*** In many ways seemingly a heterogeneous text, the Gita reconciles many facets and schools of Hindu philosophy, including those of Brahmanical (*i.e.*, orthodox Vedic) origin and the parallel ascetic and Yogic traditions. It comprises primarily Vedic (as in the four Vedas, as opposed to the Upanishads/Vedanta), Upanishadic, Sankhya and Yogic philosophies. For its religious depth, quintessential Upanishadic and Yogic philosophy and beauty of verse, the Bhagavad Gita is one of the most compelling and important texts of the Hindu tradition. It is considered by many as one of the world's greatest religious and spiritual scriptures.

It had always been a creative text for Hindu priests and Yogis. Although it is not strictly part of the 'canon' of Vedic writings, almost all Hindu traditions draw upon the Gita as authoritative. Some claim that it may have been inserted into the Mahabharata at a later date, but this is only natural as it sounds more like an Upanishad (which are commentaries that followed the Vedas) in thought than a Purana (histories), of which tradition the Mahabharata is a part.

For the Vedantic schools of Hindu philosophy, it belongs to one of the three foundational texts (Sanskrit: Prasthana Trayi, literally three points of departure) (the other two being the Upanishads and Brahma Sutras). Every such school is required to have a commentary on the three. The oldest available commentary is from Adi Shankara but he mentions older commentators. He is followed by classical commentators like Anandagiri, Shridhara Swami, Madhuskdana Sarasvati, Ramanuja, Madhvacharya, Nimbarka, Vallabha and Dnyaneshwar. While the traditional text commented upon by many scholars, including Adi Shankara and Ramanuja, consists of 700 verses, a recension of the text from Kashmir has an additional 15. The renowned philosopher Abhinavagupta (10-11th century BC) has written a commentary on this recension called Gitartha-Samgraha. Other ancient and medieval scholars (like Vedanta Desika in the Tatparya-Chandrika) seem to be aware of such additional verses but prefer not to comment on them.

Among the great sages and philosophers who have drawn inspiration from the Bhagavad Gita is Sri Chaitanya Mahaprabhu, who initiated public singing of the "Hare Krishna" mantra. Mahatma Gandhi derived great moral strength from Bhagavad gita, which is evident in his words:

"The Geeta is the universal mother. I find a solace in the Bhagavadgeeta that I miss even in the Sermon on the Mount. When disappointment stares me in the face and all alone I see not one ray of light, I go back to the Bhagavad Gita. I find a verse here and a verse there, and I immediately begin to smile in the midst of overwhelming tragedies-and my life has been full of external tragedies-and if they have left no visible or indelible scar on me, I owe it all to the teaching of Bhagavad Gita."

Swami Vivekananda, the follower of Sri Ramakrishna, was known for his commentaries on the four Yogas-Bhakti, Jnana, Karma and Raja Yoga. He drew from his knowledge of the Gita to expound on these Yogas. Swami Sivananda advises the aspiring Yogi to read verses from the Bhagavad Gita every day. Paramahamsa Yogananda, writer of the famous "Autobiography of a Yogi", viewed the Bhagavad Gita as one of the world's most divine scriptures.

A minority of Hindus, belonging to Saivite and Shakti-centered sects, reject the Gita's authority. Upon witnessing the world's first atomic blast in 1945, J. Robert Oppenheimer, American physicist and dirctor of the Manhattan Project, is reported to have quoted "I am become Death, the shatterer of worlds".

## Ramayana

The *Ramayana* is a Sanskrit epic attributed to the poet Valmiki and is an important part of the Hindu canon (*smrti*). The name *Ramayana* is a tatpurusha compound of *Rma* and *ayana* "going, advancing", translating to "the travels of *Rama*". The *Ramayana* consists of 24,000 verses in seven cantos (*knas*) and tells the story of a prince, Rama of Ayodhya, whose wife Sita is abducted by the demon (*Rkshasa*) king of Lanka, Rvana. In its current form, the Valmiki Ramayana is dated variously from the 5th century to the 1st centuries BCE, or about coeval to early versions of the *Mahabhrata*.

As with most traditional epics, since it has gone through a long process of interpolations and redactions, it is impossible to date it accurately. The Ramayana had an important influence on later Sanskrit poetry, primarily through its establishment of the Sloka meter. But, like its epic cousin Mahabhrata, the Ramayana is not just an ordinary story. It contains the teachings of ancient Hindu sages and presents them through allegory in narrative and the interspersion of the philosophical and the devotional. The characters of Rama, Sita, Lakshmana, Bharata, Hanumn and Rvana (the villain of thȩ piece) are all fundamental to the cultural consciousness of India.

One of the most important literary works on ancient India, the Ramayana has had a profound impact on art and culture in the Indian Subcontinent. The story of Rama also inspired a large amount of later-day literature in various languages, notable among which are the works of the sixteenth century Hindi poet Tulsidas and the Tamil poet Kambar of the 13th century. The Ramayana is not just a Hindu religious tale. Starting from the 8th century, the colonisation of Southeast Asia by Indians began. Several large empires like the Khmers, the Majapahits, the Saïlendras, the Champas and Sri Vijaya were established. Because of this, the Ramayana became popular in Southeast Asia and manifested itself in text, temple architecture and performance, particularly in Indonesia (Java, Sumatra and Borneo), Thailand, Cambodia, Malaysia, Vietnam and Laos.

***Structure of Valmiki Ramayana:*** Valmiki's *Ramayana*, the oldest and most widely read version of Ramayana is the basis of all the various version of Ramayana that are prevalent in the various cultures. The text survives in numerous complete and partial manuscripts, the oldest surviving of which is dated from the eleventh century BC. The current text of Valmiki Ramayana has come down to us in two regional versions from the north and the south of India. Valmiki Ramayana has been traditionally divided into seven books, dealing with the life of Rama from his birth to his death.

1. Bala Kanda – Book of the young Rama which details the miraculous birth of Rama, his early life in Ayodhya, his slaying of the demons of the forest at the request of Vishvamitra and his wedding with Sita.
2. Ayodhya Kanda – Book of Ayodhya in which Dasharatha comes to grief over his promise to Kaikeyi and the start of Rama's exile.
3. Aranya Kanda – Book of the Forest which describes Rama's life in the forest and the abduction of Sita by Ravana.
4. Kishkinda Kanda – Book of Kishkinda, the Vanara kingdom in which Rama befriends Sugriva and the Vanara army and begins the search for Sita.
5. Sundara Kanda – Book of Sundara (Hanuman) in which Hanuman travels to Lanka and finds Sita imprisoned there and brings back the good news to Rama.
6. Yuddha Kanda – Book of the War, which narrates the Rama-Ravana war and the return of the successful Rama to Ayodhya and his coronation.
7. Uttara Kanda – Epilogue, which details the life of Rama and Sita after their return to Ayodhya, Sita's banishment and the eventual demise of Sita and Rama.

There have been speculations on whether the first and the last chapters of Valmiki's *Ramayana* were indeed written by the original author. Many experts are of the opinion that they are integral part of the book in spite of the many differences in style and some contradictions in content between these two chapter and the rest of the book.

These two chapters contain most of the mythological interpolations found in the Ramayana, such as the miraculous birth of Rama and his divine nature as well as the numerous legends surrounding Ravana.

Main Characters of Ramayana:

- Rama is the hero of this epic tale. He is portrayed as an incarnation of the god Vishnu. He is the eldest and the favourite son of the King of Ayodhya, Dasharatha. He is a popular prince loved by one and all. He is the epitome of virtue. Dasaratha, forced by one of his wives Kaikeyi commands Rama to relinquish his right to the throne and go into exile by his father. While in exile, Rama kills the demon king Ravana.
- Sita is the wife of Rama and the daughter of king Janaka. Sita is the epitome of womanly purity and virtue. She follows her husband into exile and there gets abducted by Ravana. She is imprisoned in the island of Lanka by Ravana. Rama rescues her by defeating the demon king Ravana.
- Hanuman is a monkey (*Vanara*) belonging to the monkey kingdom of Kishkinda. He worships Rama and helps find Sita by going to the kingdom of Lanka crossing the great ocean.
- Lakshmana, the younger brother of Rama, chose to go into exile with him. He spend his time protecting Sita and Rama. He is deceived by Ravana and Maricha into believing that Rama was in trouble while Sita gets abducted.
- Ravana is a demon who is the king of Lanka. He received a boon from Brahma the he will not be killed by either gods, demons or by spirits, after performing a severe penance for ten thousand years. He has ten heads and twenty arms. After getting his reward from Brahma, Ravana begins to lay waste the earth and disturbs the deeds of good Brahmins. Rama is born a human to defeat him, thus overcoming the boon given by Brahma.
- Dasharatha is the king of Ayodhya and the father of Rama. He has three queens, Kousalya, Sumitra and Kaikeyi, and three other sons, Bharata, Lakshmana and Shatrughna. Kaikeyi, Dasharatha's favourite queen forces him to make his son Bharata heir apparent and

send Rama into exile. Dashatara dies heartbroken after Rama goes into exile.

- Bharata is the second son of Dasharata. When he learns that his mother Kaikeyi had forced Rama into exile and caused Dasharata to die broken hearted, he storms out of the palace and goes in search of Rama. When Rama refuses to break his exile to return to the capital to assume the throne, he requests and gets Rama's sandals and places them on the throne.
- Vishwamitra is the sage who takes Rama into the forest at the behest of defeating the demons destroying his Vedic sacrifices. On the way back he takes Rama into Mithila where Rama sees and falls in love with Sita.

***Synopsis:*** Rama, the hero of *Ramayana*, is a popular deity worshiped by Hindus, the route of his wanderings being, each year, trodden by devout pilgrims. The poem is not a mere literary monument, it is a part of Hinduism, and is held in such reverence that the mere reading or hearing of it, or certain passages of it, is believed by the Hindus to free them from sin and grant every desire to the reader or hearer. According to Hindu tradition, is an incarnation (Avatar), of the god Vishnu, who is part of the Hindu Trinity. The main purpose of this incarnation is to demonstrate the righteous path (dharma) for the life on earth.

***Rama's Youth:*** Brahma, creator of the universe, could not revoke a boon he gave the demon king Ravana, as a reward for his severe penances, that he should not be slain by gods, demons, or spirits. Having been then rewarded, Ravana began with the help of his evil supporters, the *Rakshasas*, to lay waste the earth and to do violence to the good, especially the Brahmin priests, disturbing their sacrifices.

All the gods, watching this devastation, went to Brahma to find a way to deliver themselves and the earth of this evil. Brahma went to Vishnu and conveyed the anguish of the gods and requested that Vishnu incarnate on earth as a human to destroy Ravana, as Ravana had not asked for protection against humans or beasts in his wishes to Brahma.

Meanwhile, the good king Dasharatha of Ayodhya, who had ruled over his kingdom of Kosala for a long time, was beginning to become anxious about his successor, for he had no sons to take over the kingdom after him. Taking advice from his ministers and priests Dasharatha organised a *Putrakameshti Yagna*, a sacrifice for progeny. Vishnu decided to be born as the eldest to Dasharatha and caused a divine being to emerge from the sacrificial fire. The divine being gave Dashratha a golden vessel filled with nectar and asked him to give it to his queens. Dasharatha divided it amongst his three queens, Kausalya, Sumitra and Kaikeyi. In due course they became pregnant and gave birth to four sons: Queen Kausalya gives birth to the eldest son, Rama. Bharata is born to Queen Kaikeyi, and twins, Lakshmana and Shatrughna, are born to Queen Sumitra.

The boys grew up learning the scriptures and the art of bowmanship from the sage Vasishta. One day, the sage Vishwamitra visited the kingdom and asked King Dasaratha to send Rama to protect him from demons who had been disturbing his sacrifices. Although very reluctant, Dasharatha agreed to send Rama and Lakshmana with Vishwamitra. As the brothers fulfilled their duties, Vishwamitra was pleased with them and bestowed upon them various celestial weapons.

Towards the end of their stay with Vishwamitra, Rama chanced to pass near the kingdom of Mithila and heard that its king, Janaka, had offered his peerless daughter, Sita, in marriage to the man who could bend the mighty bow of god Siva, which had been kept at Janaka's court. Rama at once determined to accomplish the feat, which had been tried in vain by so many suitors. When he presented himself at court Janaka was at once won by his youth and beauty. Five thousand men drew in the mighty bow, resting upon an eight-wheeled chariot. Rama without any apparent effort bent it until it broke and, Janaka gladly gave him his eautiful daughter. After the splendid wedding ceremonies were over, the happy pair travelled back to Ayodhya.

***Rama's Exile:*** King Dasaratha, began to feel weary of reigning, and decidd to make Rama, his eldest son and heir-apparent, the co-regent (*Yuvaraja*). His happy people received the announcement of his intention with delight and the whole city was in the midst of the most splendid preparations for the ceremony. Dasaratha went to discuss the celebrations with his favourite wife Kaikeyi. However, Kaikeyi, her jealousy aroused by her evil maid Manthara, because the son of Kausalya and not her darling Bharata, at that time absent from the city, was to be made Yuvaraja, fled to an ante-chamber where Dasharata found her in tears..

To Dasharatha's concerned queries, Kaikeyi recalled that, ages ago, the old king had granted her two boons. She now demanded the fulfilment of these before she would consent to smile upon him. Dasharatha agreed and Kaikeyi revealed her demands. She required him, first, to appoint her son Bharata as co-regent and, second, to exile Rama for fourteen years to the terrible forest of Dandaka.

Dasharatha was grief-stricken, but agreed to abide by his promise. Rama, the obedient son, immediately agreed to relinquish his claim to the throne and started to leave for his exile. His faithful wife Sita and his loving brother Lakshmana also decided to go along with Rama. With Dasharatha lying grief-stricken, Rama left for the forest, followed by the lamenting people of Ayodhya. Soon after, king Dasharatha died, overcome by grief.

***Sita's Abduction:*** Rama, Sita and Lakshmana left behind Ayodhya and its people, crossed the river Ganges and went into the forest. They found an idyllic place called Chitrakuta to establish their hermitage. No more beautiful place could be imagined. Flowers of every kind, delicious fruits, and on every side the most pleasing prospects, together with perfect love, made their hermitage a paradise on earth. In the forest, Rama befriended the old vulture-king, Jatayu.

Meanwhile, Bharata returned to Ayodhya and, being also devoted to Rama, became furious with Kaikeyi for her role in exiling Rama and for the death of his father. Determined to bring back Rama, he headed for the forest. When he found Rama and pleaded with him to return and assume the throne, Rama politely refused, saying that he was duty-bound to see that his father's promise was fulfilled. Reluctantly Bharata agreed to return to the kingdom, requesting that Rama give to him his sandals. Back in Ayodhya, Baratha placed Rama's sandals on the throne of Ayodhya, and ruled as Rama's proxy from a village called Nandigrama near Ayodhya, awaiting his return. He also vowed to end his own life if Rama failed to return after fourteen years.

One day, the demoness Surpanakha, a sister of the demon king Ravana, chanced upon Rama's hermitage and saw the handsome Rama and became enamored. Taking the form of a beautiful young girl, she tried to seduce Rama. Rama, ever faithful to his wife Sita, did not respond. But Lakshmana, infuriated by Surpanaka's act of willful lasciviousness, severed her nose and ears. Surpanakha flew back to Ravana complaining. Ravana, after hearing of the beautiful Sita from Surpanakha, resolved to kill Rama in revenge and take Sita for himself. With the help of the demon Maricha Ravana lured Rama and Lakshmana away from the hermitage, leaving Sita alone.

Before leaving Lakshmana drew a circle in the dirt saying that Sita would be safe as long as she stayed in the circle. Ravana approached the hermitage in the guise of an old man and asked Sita to give him some food. Initially hesitant to step out of Lakshmana's circle, Sita finally stepped out to give the old man some food. At this moment Ravana grabbed Sita and fled in his airborne vehicle, (Pushpaka Vimana). Jatayu, seeing them fly, attempted to save Sita, but Ravana engaged Jatayu in combat and choped off the vulture's wings. On returning to the hermitage, Rama and Lakshmana found it empty and anxiously began a search. Through Jatayu, whom they found lying mortally wounded, Rama and Lakshmana learnt of Sita's fate.

***Monkey Kingdom:*** Continuing their search, they encountered the monkey king of Kishkindha, Ṣugriva, and Hanuman, one of his generals, among whom Sita had dropped from the chariot her scarf and some ornaments. Sugriva had been deposed from his kingdom by his brother, Vali, who had also taken his wife Roma from him. Rama agreed to defeat Vali if Sugriva would assist in the search for Sita. The agreement made, Sugriva challenged Vali to a duel. While the duel was progressing, Rama shot from his bow and killed Vali. Sugriva regained his kingdom and his wife.

Sugriva and Rama sent the monkey soldiers in various directions in search of Sita. However their efforts didn't bear fruit until they met another ancient vulture, Sampati, who was the brother of the slain Jatayu. Sampati was earthbound and deformed-his wings were burnt when he flew too close to the Sun. His brother, being hardier, had saved him from falling to his death. While Jatayu was the physically stronger of the two, Sampati possessed a compensating gift of vision. Sampati's vision was incredibly powerful, spanning several hundred yojanas and enabling him to see farther than anyone else. On hearing of Ravana's killing his brother, he readily agreed to help the monkeys. He was soon able to spot Sita in the southern direction. He could see her imprisoned in a garden of Ashoka trees on the island of Lanka, beyond the southern ocean.

***Hanuman in Lanka:*** Sugriva dispatched his monkey army to the south with his nephew Angad at the head. Hanuman went with Angad as his general. Whey they reached deep south, they found a great ocean stretching between them and land of Lanka. They could find no means by which to cross the ocean. Commanding his soldiers to remain where they were, Hanuman expanded his body to enormous proportions, leaped the vast expanse of water, and alighted upon a mountain *Trikuta* from which he could look down upon Lanka. Perceiving the city to be closely guarded, he assumed the form of a cat, and thus, unsuspected, crept through the barriers and examined the city. He found Ravana in his apartments, surrounded by beautiful women, but Sita was not among them. Continuing his search, he at last discovered her, her beauty dimmed by grief, seated under a tree in a beautiful asoka grove, guarded by hideous rakshasas with the faces of buffaloes, dogs, and swine.

Assuming the form of a tiny monkey, Hanuman crept down the tree, and giving her the ring of Rama, took one from her. He offered to carry her away with him, but Sita declared that Rama must himself come to her rescue, and as proof of finding her Sita gave Hanuman a priceless jewel to take back to Rama.

While they were talking together, Ravana appeared, and, after fruitless wooing, announced that if Sita did not yield herself to him in two months he would have her guards "mince her limbs with steel" for his morning repast.

In his rage, Hanuman destroyed a mango grove and was captured by the rakshasa guards, and brought before Ravana. Hanuman proclaimed that he was a messenger of Rama, and demanded that Ravana restore Sita to Rama or fall victim to Rama's wrath. Furious at hearing Hanuman's words, Ravana ordered Hanuman's death.

Vibhishana, Ravana's righteous brother, intervened and counseled Ravana to follow the scriptures, reminding that it was improper to execute a messenger, and instead told him to exact the appropriate punishment for Hanuman's crime. Ravana accepted and ordered his rakshasas to set fire to Hanuman's tail. As soon as this was done, Hanuman made himself very small, slipped from his bonds, and, jumping upon the roofs, spread a conflagration through the city of Lanka. He leaped back to the mainland, conveyed the news of Sita's captivity to Rama and Sugriva, and was soon engaged in active preparations for the campaign.

***Battle of Lanka:*** Rama decided that as long as the ocean was not bridged, it was impossible for any one but Hanuman to cross it. Rama meditated for three days, but in vain. In his anger at being so ignored, Ram turned his weapons against the ocean, until from the terrified waves arose Varuna, the god of the ocean, who promised him that if he built a bridge, the waves should support the materials as firmly as though it were built on land.

Terror reigned in Lanka at the news of the approach of Rama. Vibishana, Ravana's brother, deserted to Rama, because of the demon's rage when he advised him to make peace with Rama. Fiercely fought battles ensued, in which even the gods took part — Vishnu and Indra taking sides with Rama, and the evil spirits fighting with Ravana.

After the war had been fought for some time, with varying results, and a great number of troops on both sides were killed, it was decided to determine the victor by single combat between Ravana and Rama. Even the gods were terrified at the fierceness of the conflict. At each shot Rama's mighty bow cut off a head of Ravana, which at once grew back, and the hero was in despair until Vibhishana told him to aim at Ravana's belly-button. Rama took careful aim.

As Ravana fell by this weapon, flowers rained from heaven upon the happy victor, and his ears were ravished with celestial music. Touched by the grief of Ravana's widow, Mandodari, Rama ordered a splendid funeral for his foe, and then sought the conquered city.

Sita was led forth, beaming with happiness at finding herself reunited to her husband; but her happiness was destined to be of short duration. Rama received her with coldness and with downcast eyes, saying that she could no longer be his wife, after having dwelt in the house of Ravana. Sita assured him of her innocence; but on his continuing to revile her, she ordred her funeral pyre to be built, since she would rather die by fire than live despised by Rama. The sympathy of all the bystanders was with Sita, but Rama saw her enter the flames without a tremor. Soon Agni, the god of fire, appeared, bearing the uninjured Sita in his arms. Her innocence thus publicly proved by the trial by fire, she was welcomed by Rama, whose treatment she tenderly forgave.

The conquest won, Ravana defeated, and Sita restored, Rama returned in triumph to Ayodhya, and assumed the governance to the great delight of Bharata and the people of Ayondhya.

***Sita Banished:*** Ayodhya was prosperous, the people were happy and for a time all went well. It was not long, however, before whispers concerning Sita's long stay in Lanka spread through the city, and Rama came to hear the whisperings that a famine in the country was due to the guilt of Sita, who had suffered the caresses of Ravana while in captivity. Under the pressure from the citizens of Ayodhya, Rama banished her to the forest in which they had spent together the happy years of their exile. Without a murmur the unhappy Sita dragged herself to the forest, and, torn with grief of body and spirit, found the hermitage of Valmiki, where she gave birth to twin sons, Lava and Kuca. Here she reared them, with the assistance of the hermit, who was their teacher, and under whose care they grew to manhood, handsome and strong.

It chanced that about the time the youths were twenty years old, Rama began to think the gods were angered with him because he had killed Ravana, who was the son of a Brahman. Rama became determined to propitiate them by means of Asvamedha, the great sacrifice, in which he caused a horse to be turned loose in the forest.

When his men went to retake it, at the end of the year, they found it caught by two strong and beautiful youths who resisted all efforts to capture them. When his men couldn't retake the horse, Rama went to the forest in person, only to learn that the youths were his twin sons, Lava and Kuca. Struck with remorse, Rama recalled the sufferings of his wife Sita, and on learning that she was at the hermitage of Valmiki, requested her to come with him.

Sita had time to recover from the love of her youth, and the prospect of life with Rama, she felt, was not altogether pleasant. She appealed to the earth, if she had never loved any man but Rama, if her truth and purity were known to the earth, let it open its bosom and take her to it. While the people stood trembling with horror, the earth opened, a gorgeous throne appeared, and the goddess of earth, seated upon it, took Sita beside her and conveyed her to the realms of eternal happiness, leaving the too late repentant people to wear out the remaining years in penitence.

***Morals in Ramayana:*** In his *Ramayana,* Valmiki expresses his view of human code of conduct through Rama: life is evanescent and the hedonistic approach to it is meaningless. However, that should not allow one to be indifferent to one's own rights and duties laid down in the ancient texts. He thus adopts the view that Dharma is what is proclaimed in the Veda and it should be followed for its own sake, not for what it brings you in pain or pleasure. Doing this will ensure one's welfare in this and the next world. In addition,

Ramayana also reinforces the need for thinking about the consequences before making promises, for if you make them you must keep them, no matter how hard it may be.

*Sankshepa Ramayana*, the brief narration of the entire Ramayana story by the age Narada to Valmiki, forms the first *sarga* of Valmiki Ramayana. Narada lists the sixteen qualities of the ideal man and says that Rama was the complete man possessing all sixteen of these qualities. Although Rama himself declares "he is but a man, and never once claims to be divine, Rama is regarded by Hindus as one of the most important Avatar of the god Vishnu and as an ideal man.

Valmiki portrays Rama not as a supernatural being, but as a human with all the attendant shortcomings, who encounters moral dilemmas but who overcomes these by simply adhering to the *dharma*—the righteous way. There are several instances narrated in Valmiki Ramayana which cast shadows on the pristine character of the hero. When Rama killed Vali to aid Sugriva regain his throne, it was not in fair combat, but while hiding behind a tree. When Sita was freed from Ravana's prison, Rama forced Sita to undergo an ordeal by fire to prove her purity and later as the king, Rama killed the Shudra Shambuka for performing a yogic penance not in keeping with his perceived low station in the society. These any many more instances within *Ramayana* illustrate the very human nature of the hero Rama help to reinforce the basic moral behind the story that men do not have to be supernatural to be righteous.

***Textual History:*** Traditionally the epic belongs to the Treta Yuga, one of the four eons of Hindu chronology and is attributed to Valmiki, is an active participant in the story.

It is composed in Epic Sanskrit, an early variant of Classical Sanskrit, so that in principle the core of the work may date to as early as the 5th century BC. Since in its current form, after hundreds of years of transmission through recitations and in manuscript form, the epic has gone thorough numerous variations, it cannot be dated by linguistic analysis as a whole, and should be considered to have emerged over a long process, spanning maybe from the 5th century BCE to the 4th century BC.

The core events told in the epic may well be of even greater age, the names of the characters, Rama, Sita, Dasharata, Janaka, Vasishta and Vishwamitra are all known in the Vedic literature such as the Brahmanas which are older than the Valmiki Ramayana. However, nowhere in the surviving Vedic poetry, is a story similar to the Ramayana of Valmiki. Brahma, one of the main characters of *Ramayana*, and Vishnu, who according to *Bala Kanda* was incarnated as Rama are not Vedic deities, and come first into prominence with the epics themselves (and further during the 'Puranic' period of the later 1st millennium BC).

There is general consensus that books two to six form the oldest portion of the epic while the first book *Bala Kanda* and the last the *Uttara Kanda* are later additions. The author or authors of *Bala Kanda* and *Ayodhya Kanda* appear to be familiar with the eastern Gangetic basin region of northern India and the Kosala and Magadha region during the period of the sixteen janapadas as the geographical and geopolitical data is in keeping with what is known about the region. However when the story moves to the *Araya Kanda* and beyond, it seems to turn abruptly into fantasy with its demn-slaying hero and fantastic creatures.

The geography of central and south India is increasingly vaguely described. The knowledge of the location of the island of Sri Lanka also lacks detail. Basing his assumption on these features, the historian H.D. Sankalia has proposed a date of the 4th century BCE for the composition of the text. A. L. Basham, however, is of the opinion that Rama may have been a minor chief who lived in the 8th or the 7th century BCE.

The events of the epic have also been dated to as early as 6000 BCE by adherents of archaeoastronomy.

***Variant Versions:*** The epic story of Ramayana was adopted by several cultures across Asia. Shown here is a Thai historic artwork depicting the battle which took place between Rama and Ravana. As in many oral epics, multiple versions of the Ramayana survive. In particular, the Ramayana related in north India differs in important respects from that preserved in South India and the rest of South-East Asia. There is an extensive tradition of oral storytelling based on the Ramayana in Thailand, Cambodia, Malayasia, Laos, Vietnam, and Indonesia. In many Malay versions, Lakshmana is given greater importance than Rama, whose character is considered somewhat weak.

***Within India:*** There are diverse regional versions of the Ramayana written by various authors in India. Some of them differ significantly from each other. During the twelfth century CE, Kamban wrote *Ramavatharam*, known popularly as Kambaramayanam in Tamil. Although based on Valmiki Ramayana, Kambaramayanam is a true classic and unique in that Kamban has modified and reinterpreted many anecdotes in Valmiki Ramayana to suit the Tamil Culture and his own ideas. Valmiki's Ramayana inspired the Sri Ramacharit Manas by Tulasidas in 1576, an epic Hindi version with a slant more grounded in a different realm of Hindu literature, that of bhakti.

It is an acknowledged masterpiece of India. It is popularly known as *Tulsi-krita Ramayana*. Gujarati poet Premanand wrote a version of *Ramayana* in the 17th century. Other versions include, a Bengali version by Krittivas in the 14th century, in Oriya by Balarama Das in the 16th century, in Marathi by Sridhara in the 18th century, a Telugu version by Ranganatha in the 15th century, a Kannada Ramayana by the 16th century poet Narahari, Kotha Ramayana in Assamese by the 14th century poet Madhava Kandali and Adhyathma Ramayanam Kilippattu, a Malayalam version by Thunchaththu Ezhuthachan in the 17th century.

There is a sub-plot to Ramayana, prevalent in some parts of India relates to the adventures of Ahi Ravana and Mahi Ravana, the evil brother of Ravana, which enhances the role of Hanuman in the story. Hanuman rescues Rama and Lakshmana after they are kidnapped by the Ahimahi Ravana at the behest of Ravana and held prisoner in a subterranean cave, ready to be sacrificed to the goddess Kali.

There have been reports of a version of the Ramayana story prevalent amongst the Mappilas of Kerala. This version, known as *Mappila Ramayana*, forms a part of the Mappillapattu. *Mappillapattu* is a genre of folk singing popular amonst the musims of Kerala and Lakshadweep. Being of Muslim origin, the hero of this story is a sultan. There are no major changes in the names of characters except for that of Rama's which is changed to 'Laman'. The language and the imagery projected in the Mappilapattu are in accordance with the social fabric of the earlier Muslim community.

***South-east Asian Versions:*** Many other Asian cultures have adapted the Ramayana, resulting in other national epics. Aspects of the Chinese epic Journey to the West was inspired by the Ramayana, particularly the character Sun Wukong, who is believed to have been based on Hanuman. Kakawin Rmayana is an old Javanese rendering of the Sanskrit Ramayana from the ninth century Indonesia. It is a faithful rendering of the Hindu epic with very little variation. Phra Lak Phra Lam is a Lao language version, whose title comes from Lakshmana and Rama. The story of Lakshmana and Rama is told as the previous life of the Buddha. In Hikayat Seri Rama of Malaysia, Dasharatha is the great-grandson of the Prophet Adam. Ravana receives boons from Allah instead of Brahma.

Thailand's popular national epic Ramakien is derived from the Hindu epic. In *Ramkien*, Sita is the daughter of Ravana and Mandodari the astrologer brother of Ravana, predicts calamity from the horoscope of Sita. So Ravana has her thrown into the waters, who, later, is picked by Janaka (*Janok*). While the main story is identical to that of the *Ramayana*, many other aspects were transposed into a Thai context, such as the clothes, weapons, topography, and elements of nature, which are described as being Thai in style. It has an expanded role for Hanuman and he is portrayed as a lascivious character. Ramakien can be seen in an elaborate illustration at the Wat Phra Kaew temple in Bangkok.

Other Southeast Asian adaptations include Ramakavaca of Bali, Maradia Lawana of the Philippines, the Reamker of Cambodia and the Yama Zatdaw of Myanmar.

***Contemporary Versions:*** Contemporary versions of the Ramayana include *Shri Ramayana Darshanam* by Dr. K. V. Puttappa (Kuvempu) in Kannada and *Ramayana Kalpavrikshamu* by Viswanatha Satyanarayana in Telugu, both of which have been awarded the Jnanpith Award. The modern Indian author Ashok Banker has so far written a series of six English language novels based on the Ramayana.

The Gaiety Railway Theatre Company of Kanchipuram has presented revisionist views of this epic, seeking to restore the pride of the Dravidian people. These versions of the Ramayana diverge sharply from traditional presentations, with Ravana portrayed as a statesman scholar with whom Sita was infatuated, and Rama as a playboy prince with little application or dedication, who ordered massive genocide on a drunken whim. Such characterisations are part of a latent and increasing attempt by Dravidian movement to reassert their culture and independence.

***Ramayana Landmarks:*** Amongst the ruins of the Vijayanagara empire near Hampi, is a cave known as Sugriva's Cave. The cave is marked by coloured markings. The place holds its similarity to the descriptions of 'kishkinda' in Sundarakanda. Rama is said to have met Hanuman here. The place is also home to the famous Hazararama temple (Temple of a thousand Ramas).

## Puranas

*Purana* is the name of a genre (or a group of related genres) of Indian written literature (as distinct from oral literature). Its general themes are history, tradition and religion. It is usually written in the form of stories related by one person to another.

There are many texts designated as 'Purana.' The most important are:

- *Mahpurnas* and *Upapurnas*-Written in Sanskrit, by Brahmins. They usually adhere to the 'Panca-laksana' typology (see below). They tell of the creation of the universe and the human race, narrate the genealogies of kings and saints, and contain assorted narratives, stories, and philosophical and religious topics.

Included among the texts of distinctly lesser importance that also bear the name Puranas are:

- *Sthala Purnas*-Scriptures usually extolling the virtues of a certain Hindu temple. They narrate stories of the temple's creation and spiritual history.
- *Kula Purnas*-Scriptures that deal with the origin and legends of a particular caste.

The designation 'Purana' implicitly confers on a text a kind of factual aura, somewhat similar to the designation 'History' in Western discourse.

***Mahapuranas and Upapuranas:*** Mah-(Sanskrit 'great', 'mighty') and Upa-(Sanskrit 'lower', 'additional') Puranas are written in Sanskrit by Brahmins. Traditionally they are said to narrate five subjects, called Panca-laksana (Sanskrit 'five distinguishing marks'), which are:

1. Sarga-The creation of the universe.
2. Pratisarga-Secondary creations, mostly recreations after dissolution.
3. Vama-Genealogy of gods and sages.
4. Manvantara-The creation of the human race and the first human beings.
5. Vamanucaritam-Dynastic histories.

Most Mahapuranas and Upapuranas deal with these subject matters, although the bulk of their text consists of historical and religious narratives. Some scholars have suggested that these 'distinguishing marks' are shared by other traditional religious scriptures of the world (*e.g.* the Bible). A Purana usually gives prominence to a certain deity — (Shiva, Vishnu or Krishna, Durga) — and depicts the other gods as subservient. Most use an abundance of religious and philosophical concepts in their narration, from Bhakti to Samkhya.

***Classification:*** Traditionally it is said that there are eighteen Mahapuranas and eighteen Upapuranas. Of the Mahapuranas it is said that six belong to the quality (*guna*) of goodness, six to passion, and six to ignorance. According to the *Padma Purana,* these are the Mahapuranas and their corresponding qualities:

- *Goodness*: Vishnu Purana, Bhagavata Purana,

Naradeya Purana, Garuda Purana, Padma Purana, Varaha Purana

- *Passion:* Brahmanda Purana, Brahma Vaivarta Purana, Markandeya Purana, Bhavishya Purana, Vamana Purana, Brahma Purana
- *Ignorance:* Matsya Purana, Kurma purana, Linga Purana, Shiva Purana, Skanda Purana, Agni Purana

Some Upapuranas are: Sanat-kumara, Narasimha, Brihan-naradiya, Siva-rahasya, Durvasa, Kapila, Vamana, Bhargava, Varuna, Kalika, Samba, Nandi, Surya, Parasara, Vasishtha, Devi-Bhagavata, Ganesa, and Hamsa. Most of these have not been critically edited yet, and are available mostly through devotional publications, in multiple versions and recensions. Apart from the system of classification above mentioned, the Puranas are enumerated and classified in multiple ways. The canonical lists of Puranas vary from place to place and from time to time.

***Author and Disemmination:*** Traditionally the Puranas are said to have been composed by the sage Vyasa, the narrator of the Mahabharata epic. Vyasa in Sanskrit means 'Divider,' and some scholars therefore take this simply as a term designating 'Editor'. The texts, these scholars say, were probably written all over India and are being rewritten and reedited to the present day.

The Puranas, being in Sanskrit, are usually not directly accessible to the common man as read texts. They are, however, available in vernacular translations and are disseminated by Brahmin scholars, who read from them and tell their stories, usually in Katha sessions (in which a travelling brahmin settles for a few weeks in a temple and narrates parts of a Purana, usually with a Bhakti perspective).

***Chronology:*** The Itihasa-Purana is mentioned in the Chandogya Upanishad. There are also references to the Puranas in the Atharvaveda 11.7.24 and the Satapatha Brahmana 11.5.6.8. and 13.4.3.13. There are probably also references to the Itihasa in the Nirukta. The Itihasa-Purana was considered to be almost as sacred and ancient as the Vedas. However, the ancient Itihasa-Purana was probably different from the later Puranas. Scholars argue that some contents of the Puranas may date to an earlier period.

***Notable Puranas:*** *The Bhagavata Purana*-It is concerned with Vishnu Bhakti, telling of the exploits and deeds of Vishnu's Avataras. Its tenth canto (its longest) narrates the deeds of Krishna and, probably for the first time in Sanskrit, tells of his exploits as a child, a theme later elaborated by many Bhakti movements.

- *The Devi-bhagavata Purana*-an Upapurana extolling the virtues of the goddess Durga as the supreme being. It has become (along with the Devi Mahatmya of the Mrkandeya Purana) a basic text for Devi worshipers.
- *The Skanda Purana*-probably the longest of all. A vast storehouse of parables, legends and stories, with multiple versions and recensions. Many untraced quotes from a Purana are conveniently attributed to this Purana.
- *The Bhavishya Purana*-Apart from its other content, it is said to contain references to the Judeo-Christian creation myth (Adam and Eve are referred to as Adama and Havyavati), the Judeo-Christian flood myth, Moses, Jesus (referred to as a-'God'), Mohammad (referred to as Mahmada-'the great intoxicator'), Queen Victoria (as Queen Vicyavati), etc. It also contains references to a certain Demon language (apparently English) containing words such as 'February'.

***Sthala Puranas:*** This corpus of texts narrates the virtues and stories connected with a certain temple or shrine (the word 'Sthala' means 'Place' in Sanskrit). There are numerous Sthala Puranas, most written in vernaculars, some with Sanskrit versions as well. Most claim to have a Sanskrit origin, and some of the Sanskrit versions also appear in a Mahapurana or an Upapurana. Some Tamil Sthala Puranas have been researched by David Dean Shulman.

***Kula Puranas:*** These are mostly Caste Puranas (the word 'Kula' means 'Family' or 'Tribe' in Sanskrit). They deal with a caste's origin myth, stories and legends. The caste purana is an important source for caste identity and is usually contested by other, rival, castes. This subgenre is usually in the vernacular and might at times be oral. This subgenre has been little researched. But it is rather well documented in the Caste section of the British Census of India Report and the various Gazzeteers.

***Other Puranas:*** There are many other narratives that go by the name of Purana. Most are written in vernaculars and usually are concerned with mythical and historical narrations. These texts, such as the Padma Purana of Bengal and Assam (narrating the story of the goddess Manas), are vast in number and scattered all over the Indian subcontinent.

***Non Hindu Puranas:*** There are many Jain Puranas, dealing with Jain myths, history and legends. Studies and translations of this particular genre are meagre. Arguably, some Buddhist Mahyna Sktras seem to have some characteristics of Puranas

***Tantras:*** The Tantra refer to numerous and varied scriptures pertaining to any of several esoteric traditions rooted in Hindu and Buddhist philosophy. Most tantra were written between the 10th and 14th centuries BC. Tantra include a variety of mystical, occult, medical and scientific works. While Hinduism is typically viewed as being Vedic, the Tantra are not considered part of the orthodox Hindu/Vedic scriptures. Legend ascribes the origin of Tantra to Dattatreya, a semi-mythological yogi and the assumed author of the Jivanmukta Gita ("Song of the liberated soul"). Others traditionally confer authorship on the Indian god Shiva. Matsyendranath is credited with authorship of the Kaulajnana-nirnaya, a voluminous ninth-century tantra dealing with a host of mystical and magical subjects, and occupies an important position in the Hindu tantric lineage, as well as in Tibetan Vajrayana Buddhism.

Buddhist and Hindu Tantra, though having many similarities from the outside, do have some clear distinctions. Scholars are unable to determine whether the Hindu or the Buddhist version of Tantra appeared first in history. Buddhist

Tantra is always part of the Mahayana school of Buddhism. Buddhist Tantra spread out from (North) India, chiefly to Tibet, where it became known as the Vajrayana school of Buddhism. It also had some influence on Chinese and Japanese Buddhism (notably Shingon).

***Sutras:*** Sktra (Sanskrit) or Sutta (Pli) literally means a rope or thread that holds things together, and more metaphorically refers to an aphorism (or line, rule, formula), or a collection of such aphorisms in the form of a manual. It is derived from the verbal root siv, meaning to sew (these words, including English to sew and Latinate suture, all derive from PIE). In Hinduism the 'sutras' form a school of Vedic study, related to and somewhat later than the Upanishads. They served and continue to act as grand treatises on various schools of Hindu Philosophy. They elaborate in succinct verse, sometimes esoteric, Hindu views of metaphysics, cosmogony, the human condition, moksha (liberation), and how to maintain a blissful, dharmic life, in a cosmic spin of karma, reincarnation and desire.

In Buddhism, the term "sutra" refers generally to canonical scriptures that are regarded as records of the oral teachings of Gautama Buddha. In Chinese, these are known as *ching*. These teachings are assembled in the second part of the Tripitaka which is called Sutra Pitaka. There are also some Buddhist texts, such as the Platform Sutra, that are called sutras despite being attributed to much later authors. The Pali form of the word, sutta is used exclusively to refer to Buddhist scriptures, particularly those of the Pali Canon.

Below is a list of sutras, organized alphabetically under the broad categories of Hinduism and Buddhism.

Sutras primarily associated with Hinduism.

*Vedanga:*

- Shiksha (phonetics)
- Chandas (metrics)
- Vyakarana (grammar)
- Ashtadhyayi (Panini), discussing grammar

- Nirukta (etymology)
- Jyotisha (astronomy)
- Kalpa (ritual)
- Srauta Sutras, performance of sacrifices
- Smarta Sutras
- Grhya Sutras, covering domestic life
- Śamayacarika or Dharma Sutras
- Sulba Sutras, architecture of sacrificial area

***Vedanta***

- Brahma Sutras (or Vedanta Sutra) (Badarayana)

***Hindu philosophy***

- Yoga Sutras; (Patanjali)
- Nyaya Sutras, on logic
- Vaisesika Sutra

Sutras primarily associated with Buddhism

***Pali Sutras***

***Tripitaka***

- Digha Nikaya: 34 sutras, including
- Brahmajala Sutta, a critique of 64 'wrong views' current at the time of the Buddha
- Mahaparinibbana Sutta, the last days of the Buddha, Theravada version
- Majjhima Nikaya: 152 sutras, including
- Aggi-Vacchagotta Sutta, on questions that lead to a thicket of views and on the extinguished fire as a metaphor for nirvana
- Anapanasati Sutta, a discourse on the method of being mindful of the breath
- Satipatthana Sutta
- Samyutta Nikaya: 2,889 sutras, including
- Dhammacakkappavattana Sutta, the Buddha's first discourse on the nature of his enlightenment
- Upanisa Sutta, on the two types of conditionality

- Anguttara Nikaya: 2,308 sutras, including
- Kalama Sutta, the Buddha's advice to the Kalamas on how to decide which practices are correct
- Metta Sutta, a rhapsody on the quality of universal loving-kindness
- Khuddaka Nikaya: 266 sutras and miscellaneous texts, including
- Dhammapada, a collection of sayings and aphorisms thought to be amongst the oldest texts

***Mahayana Sutras***

- Amitabha Sutra, or *Smaller Pure Land Sutra*
- Avatamsaka Sutra or *Flower Garland Sutra* (Chinese: *Huayan Jing*)
- Contemplation Sutra, important in Pure Land Buddhism
- Diamond Sutra
- Heart Sutra, important in Zen Buddhism
- Infinite Life Sutra, or *Large Pure Land Sutra*
- Lankavatara Sutra
- Lotus Sutra, foundational text of Tiantai and Nichiren sects
- Mahaparinirvana Sutra, the last days of the Buddha, Mahayana version
- Perfection of Wisdom or *Prajnparamita Sutras*
- Platform Sutra of the Sixth Patriarch
- Sutra of Perfect Enlightenment (Chinese: *Yuanjue Jing*)
- Shurangama Sutra
- Sutra in Forty-Two Sections
- Sutra of Golden Light, important in Japanese Buddhism
- Ten Stages Sutra
- The Blessing of the Original Vow of the Master of Healing, Azure Radiance Tathagata
- Ullambana Sutra
- Vimalakirti-nirdesa Sutra

***Vajrayana Sutras***

- Mahavairocana Sutra also known as the Mahavairocandhisambodhi Tantra.
- Vajrasekhara Sutra

***Other Sutras***

- Kama Sutra, written by Vatsyayana, the sutra of kama (sensual gratification), explains sexual positions.
- Smokey the Bear Sutra, Written by American poet Gary Snyder (b. 1930) in 1969, which presents 20th century environmental concerns and convictions in the form of a Buddhist sutra. It is widely available on the internet due to the author's grant of free reproduction of the text. See Wikisource-Smokey the Bear Sutra.

## Stotras

Stotras are Hindu prayers that praise aspects of God, such as Devi, Siva, or Vishnu. Stotras, according to Swami Tapasyananda, are invariably uttered aloud and consist of chanting verses conveying the glory and attributes of God. The three most famous and sacred stotras in Hinduism are Lalitha Sahasranama, Shri Rudram Chamakam, and Vishnu Sahasranama. Other famous Stotras are: Sri Ramraksha Stotra, Sri Maruti Stotra.

## Nyaya

Nyaya is the name given to one of the six orthodox or *astika* schools of Hindu philosophy — specifically the school of logic. The Nyaya school of philosophical speculation is based on texts known as the Nyaya Sutras, which were written by Aksapada Gautama from around the 6th century BC.

***Overview:*** The most important contribution made by the Nyaya school to modern Hindu thought is its methodology. This methodology is based on a system of logic that, subsequently, has been adopted by the majority of the other Indian schools, orthodox or not. This is comparable to how Western science and philosophy can be said to be largely based on Aristotelian logic.

However, Nyaya differs from Aristotelian logic in that it is more than logic in its own right. Its followers believed that obtaining valid knowledge was the only way to obtain release from suffering. They therefore took great pains to identify valid sources of knowledge and to distinguish these from mere false opinions.

According to the Nyaya school, there are exactly four sources of knowledge (pramanas): perception, inference, comparison and testimony. Knowledge obtained through each of these can, of course, still be either valid or invalid. As a result, Nyaya scholars again went to great pains to identify, in each case, what it took to make knowledge valid, in the process creating a number of explanatory schemes. In this sense, Nyaya is probably the closest Indian equivalent to contemporary analytic philosophy.

***Epistemology:*** The Naiyanikas (the Nyaya scholars) accepted four means of obtaining knowledge (*pramana*), *viz.*, Perception, Inference, Comparison and Word.

- *Perception*, called *Pratyakaha*, occupies the foremost position in the Nyaya epistemology. Perception is defined by sense-object contact and is unerring. Perception can be of two types:
- Ordinary (*Laukika* or *Sadharana*), of six types, *viz.*, visual-by eyes, olfactory-by nose, auditory-by ears, tactile-by skin, gustatory-by tongue and mental-by mind.
- Extra-ordinary (*Alaukika* or *Asadharana*), of three types, *viz.*, *Samanyalakshana* (perceiving generality from a particular object), *Jnanalakshana* (when one sense organ can also perceive qualities not attributable to it, as when seeing a chili, one knows that it would be bitter or hot), and *Yogaja* (when certain human beings, from the power of *Yoga*, can perceive past, present and future and have supernatural abilities, either complete or some). Also, there aretwo modes or steps in perception, *viz.*, *Nirvikalpa*, when one just perceives an object without being able to know its featues, and *Savikalpa*, when one is able to clearly

know an object. All laukika and alaukika pratyakshas are savikalpa. There is yet another stage called *Pratyabhijn*, when one is able to rerecognise something on the basis of memory.

- Inference, called *Anumana*, is one of the most important contributions of Nyaya. It can be of two types-inference for oneself (*Svarthanumana*, where one does not need any formal procedure, and at the most the last three of their 5 steps), and inference for others (*Parathanumana*, which requires a systematic methodology of 5 steps). Inference can also be classified into 3 types: *Purvavat* (infering an unperceived effect from a perceived cause), *Sheshavat* (infering an unperceived cause from a perceived effect) and *Samanyatodrishta* (when inference is not based on causation but on uniformity of coexistence). A detailed anaysis of error is also given, explaining when anumana could be false.
- Comparison, which is the rough transplation of *Upamana*. It is the knowledge of the relationship between a word and the object denoted by the word. It is produced by the knowledge of resemblance or similarity, given some pre-description of the new object beforehand.
- Word, or *Shabda* are also accepted as a *pramana*. It can be of two types, *Vaidika* (Vedic), which are the words of the four sacred Vedas, and are described as the Word of God, having been composed by God, and *Laukika*, or words and writings of trustworthy human beings.

***Theory of Inference:*** The methodology of inference involves a combination of induction and deduction by moving from particular to particular via generality. It has five steps, as in the example shown:

- There is fire on the hill (called *Pratijn*, required to be proved)
- Because there is smoke there (called *Hetu*, reason)

- Wherever there is fire, there is smoke (called *Udaharana*, *i.e.*, example)
- There is smoke on the hill (called *Upanaya*, reaffirmation)
- Therefore there is fire on the hill (called *Nigamana*, conclusion)

In Nyaya terminology for this example, the hill would be called as *paksha* (minor term), the fire is called as *sadhya* (major term), the smoke is called as *hetu*, and the relationship between the smoke and the fire is called as *vyapti* (middle term). Hetu further has five characteristics: (1) It must be present in the Paksha, (2) It must be present in all positive instances, (3) It must be absent in all negetive instances, (4) It must not incompatible with the minor term or Paksha and (5) All other contradictions by other means of knowledge should be absent. The fallacies in Anumana (*hetvbhasa*) may occur due to the following:

- *Asiddha*: It is the unproved hetu that results in this fallacy. [Paksadharmata]
- *shrayasiddha*: If Paksha [minor term] itself is unreal, then there cannot be locus of the hetu. *e.g.* The sky-lotus is fragrant, because it is a lotus like any other lotus.
- *Svarupasiddha*: Hetu cannot exist in paksa at all. *E.g.* Sound is a quality, because it is visible.
- *Vyapyatvasiddha*: Conditional hetu. 'Wherever there is fire, there is smoke'. The presence of smoke is due to wet fuel.
- *Savyabhichara*: This is the fallacy of irregular hetu.
- *Sadharana*: The hetu is too wide. It is present in both sapaksa and vipaksa. 'The hill has fire because it is knowable'.
- *Asadharana*: The hetu is too narrow. It is only present in the Paksha, it is not present in the Sapaksa and in the Vipaksha. 'Sound is eternal becase it is udible'.
- *Anupasamhari*: Here the hetu is non-exclusive. The hetu is all-inclusive and leaves nothing by way of

sapaksha or vipaksha. *e.g.* 'All things are non-ternal, because they are knowable'.

- *Satpratipaksa*: Here the hetu is contradicted by another hetu. If both have equal force, then nothing follows. 'Sound is eternal, because it is audible', and 'Sound is non-eternal, because it is produced'. Here 'audible' is counter-balanced by 'produced' and both are of equal force.
- *Badhita*: When another proof (as by perception) definitely contradicts and disproves the middle term (hetu). 'Fire is cold because it is a substance'.
- *Viruddha*: Instead of proving something it is proving the opposite. 'Sound is eternal because it is produced'.

***The Nyaya Theory of Causation:*** A *cause* is defined as an unconditional and invariable antecedent of an *effect* and an effect as an unconditional and invariable consequent of a cause. The same cause produces the same effect; and the same effect is produced by the same cause. The cause is *not* present in any hidden form whatsoever in its effect.

The following conditions should be met:

[1] The cause must be antencedent [Purvavrtti]

[2] Invariability [Niyatapurvavrtti]

[3] Unconditionality [Ananyathasidha]

Nyaya recognizes five kinds of accidental antecedents [Anyathasiddha]

[a] Mere accidental antecedent. *E.g.*, The colour of the potter's cloth.

[b] Remote cause is not a cause because it is not unconditional. *E.g.*, The father of the potter.

[c] The co-effects of a cause are not causally related.

[d] Eternal substances, or eternal conditions are not unconditional antecedents. *e.g.* space.

[e] Unnecessary things, *e.g.* the donkey of the potter.

Nyaya recognizes three kinds of cause:

[1] Samavayi-material cause. *E.g.* Thread of a cloth.

[2] Asamavayi-Colour of the thread which gives the colour of the cloth.

[3] Nimitta-Efficient cause, *e.g.* the weaver of the cloth.

***Anyathakyativada of Nyaya:*** The Nyaya theory of error is similar to that of Kumarila's Viparita-khyati (see Mimamsa). The Nyayayikas also believe like Kumarila that error is due to a wrong synthesis of the presented and the represented objects. The represented object is confused with the presented one. The word 'anyatha' means 'elsewise' and 'elsewhere' and both these meanings are brought out in error. The presented object is perceived elsewise and the represented object exists elsewhere. They further maintain that knowledge is not intrinsically valid but becomes so on account of extraneous conditions (*paratah pramana* during both validity and invalidity).

***Nyaya Argument for the Existence of God:*** Early Naiyanikas wrote very little about God, *i.e.*, Ishvara (literally, the Supreme Lord). However, later Buddhists in India had become from agnostic to strictly atheistic. As a reaction, the later Naiyanikas entered into disputes with the Buddhists and tried to prove the existence of God through logic. They made this question a challenge to their own existence. They gave the following nine proofs for the existence of God, enlisted in Udayana's *Nyaya Kusumanjali*:

- *Kryt* (lit. "from effect"): An effect is produced by a cause, and similarly, the universe must also have a cause. Causes (according to Naiyanikas) are of three kinds-Samavayi (in case of the universe, the atoms), Asamavayi (the assosiation of atoms) and Nimitta (which is Ishvara). The active cause of the world must have an absolute knowledge of all the material of creation, and hence it must be God. Hence from the creation, the existence of the Creator is proved.
- *yojant* (lit., from combination): Atoms are inactive and properties are unphysical. So it must be God who creates the world with his will by causing the atoms to join. Self-combination of inanimate and lifeless things is not possible, otherwise atoms would only combine at

random, creating chaos. There is to be seen the hand of a wise organiser behind the systematic grouping of the ultimate atoms into dyads and molecules. That final organiser is God.

- *Dhwite* (lit., from support): Just as a material thing falls off without a support, similarly, God is the supporter and bearer of this world, without which the world would not have remained integrated. This universe is hence superintended within God, which proves his existence.
- *Padt* (lit., from word): Every word has the capability to represent a certain object. It is the will of God that a thing should be represented by a certain word. Similarly, no knowledge can come to us of the different things here, unless there is a source of this knowledge. The origin of all knowledge should be omniscient, and, consequently, omnipotent. Such a being is not to be seen in this universe, and so it must be outside it. This being is God.
- *Pratyatah* (lit, from faith): the Hindu holy scriptures, the Vedas, are regarded as the source of eternal knowledge. Their knowledge is free from fallacies and are widely believed as a source of proof. Their authors cannot be human beings because human knowledge is limited. They cannot obtain knowledge of past, present and future and indepth knowledge of mind. Hence only God can be the creator of the Vedas. Hence his existence is proved from his being the author of the Vedas, which he revealed to various sages over a period of time.
- *Shruteh* (lit., from scriptures): The Shrutis, *i.e..*, the Vedas extol God and talk about his existence. "He is the lord of all subjects, omniscient and knower of one's internal feelings; He is the creator, cause and destroyer of the world", say the Shrutis. The Shrutis are regarded as a source of proofs by Naiyanikas. Hence the existence of God is proved.
- *Vkyt* (lit., from precepts): Again, the Veda must have been produced by a person because it has the nature

of "sentences,", in other words, the sentences of the Veda were produced by a person because they have the nature of sentences, just as the sentences of beings like ourselves. That person must have been God.

- *Samkhyvi hesht* (lit., from the speciality of numbers): The size of a dyad or a molecule depends on the number of the atoms that go to constitute it. This requisite number of the atoms that go to form a particular compound could not have been originally the object of the perception of any human being; so its contemplator must be God.
- *Adwihct* (lit., from the unforseen): It is seen that some people in this world are happy, some are in misery. Some are rich and some poor. The Naiyanikas explain this by the concept of Karma and reincarnation. The fruit of an individual's actions does not always lie within the reach of the individual who is the agent. There ought to be, therefore, a dispenser of the fruits of actions, and this supreme dispenser is God.

***Nyaya Arguments for Monotheism:*** Not only have the Naiyanikas given proofs for the existence of God, but they have also given an argument that such a God can only be one. In the *Nyaya Kusumanjali*, this is discussed against the proposition of the Mimamsa school—that let us assume there were many demigods (Devas) and sages (rishis) in the beginning, who wrote the Vedas and created the world. Nyaya says that:

[if they assume such] omniscient beings, those endowed with the various superhuman faculties of assuming infinitesimal size, etc., and capable of creating everything, then we reply that the *law of parsimony* bids us assume only one such, namely Him, the adorable Lord. There can be no confidence in a non-eternal and non omniscient being, and hence it follows that according to the system which rejects God, the tradition of the Veda is simultaneously overthrown; there is no other way open.

In other words, Nyaya says that the polytheist would have to give elaborate proofs for the existence and origin of his several celestial spirits, none of which would be logical. So it is much more logical to assume only One, eternal and omniscient God.

***Vaisheshika:*** Vaisheshika, also Vaisesika, is one of the six Hindu schools of philosophy (orthodox Vedic systems) of India. Historically, it has been closely associated with the Hindu school of logic, Nyaya. Vaisesika espouses a form of atomism and postulates that all objects in the physical universe are reducible to a finite number of atom. Originally proposed by the sage Kanda (or *Kana-bhuk*, literally, *atom-eater*) from around the 6th century BC.

***Overview:*** Although the Vaishesika system developed independently from the Nyaya, the two eventually merged because of their closely related metaphysical theories. In its classical form, however, the Vaishesika school differed from the Nyaya in one crucial respect: where Nyaya accepted four sources of valid knowledge, the Vaishesika accepted only perception and inference. Vaishesika atomism also differs from the atomic theory of modern science: according to the Vaishesikas, the functioning of atoms was guided or directed by the will of the Supreme Being. This is therefore a theistic form of atomism.

An alternative view would qualify the above in that the holism evident in the ancient texts mandate the identification of six separate traditional environments of philosophy, consisting of three sets of two pairs.

***The Logic of Vaisheshika:*** Being associated with the school of logic, early Vaisesika texts use logical argument to prove that the world is made of indivisible atoms: Assume that the world is not made of indivisible atoms, and that matter is continuous. Take a stone. One can divide this up into infinitely many pieces (since matter is continuous). Now, the Himalayan mountain range also has infinitely many pieces, so one may build another Himalayan mountain range with the infinite number of pieces that one has. One begins with a stone and ends up with the Himalayas, which is obviously ridiculous-so the original assumption that matter is continuous must be wrong, and so all objects must be made up of a finite number of pieces, called *atoms*. Later Vaisesika philosophers tried to develop a theory to explain the properties of materials as the interaction of different types of atoms that make up the material. However, Vaisesika philosophy states that these interactions are driven by divine will, making this school of philosophy a theistic one. Epistemologically, the Vaisesika school accepts perception (*pratyaksha*) and inference (*anumna*) as valid sources of knowledge.

Later Vaisesika commentators include *Prasastapada* and *Sridhara*. Over the centuries, the school merged with the Nyaya system of Indian philosophy to form the combined school of *Nyaya-Vaisesika*. The school suffered a natural decline in India after the 15th century.

## Samkhya

Samkhya, also Sankhya, is one of the schools of Indian philosophy. It is one of the six astika (that which recognizes vedic authority) systems of Hindu philosophy. It is regarded as the oldest of the orthodox philosophical systems in Hinduism, predating Buddhism of circa 500 BCE. Its philosophy regards the universe as consisting of two eternal realities: Purusha and Prakrti; it is therefore a strongly dualist and enumerationist philosophy. The Purusha is the centre of consciousness, whereas the Prakriti is the source of all material existence.

The Sankhya school has deeply influenced the Hindu Yoga school of philosophy. They are sometimes referred together as Samkhya-yoga school. Sage Kapila is traditionally considered to be the founder of the Sankhya school, although no historical verification is possible. The definitive text of classical Sankhya is the extant *Sankhya Karika*, written by *Ishvara Krishna*, circa 200 CE.

***Epistemology of Sankhya:*** According to the Sankhya school, knowledge is possible through three *pramanas* (means of knowledge)-

- *Pratyaksha*-direct sense perception
- *Anumana*-logical inference
- *Sabda*-verbal testimony

***Metaphysics of Samkhya:*** Metaphysically, Samkhya maintains a radical duality between spirit/consciousness (*Purusha*) and matter (*Prakrti*). All physical events are considered to be manifestations of the evolution of *Prakrti*, or primal Nature (from which all physical bodies are derived). Each sentient being is a *Purusha*, and is limitless and unrestricted by its physical body. Samsaara or bondage arises when the *Purusha* does not have the discriminate knowledge and so is misled as to its own identity, confusing itself with the physical body-which is actually an evolute of *Prakriti*. The spirit is liberated when the discriminate knowledge of the difference between conscious Purusha and unconscious Prakriti is realized.

The most notable feature of Sankhya is its unique theory of Cosmic evolution (not connected with Darwin's evolution). Sankhya theorizes that Prakriti is the source of the world of becoming. It is pure potentiality that evolves itself successively into twenty four tattvas or principles. The evolution itself is possible because *Prakriti* is always in a state of tension among its constituent strands-

- *Sattva*-a template of balance or equilibrium;
- *Rajas*-a template of expansion or activity;
- *Tamas*-a template of inertia or resistance to action.

All macrocosmic and microcosmic creation uses these templates. The twenty four principles that evolves are-

- *Prakriti*-The most subtle potentiality that is behind whatever that is created in the physical universe.
- *Mahat*-first product of evolution from Prakriti, pure potentiality. *Mahat* is also considered to be the principle responsible for the rise of buddhi or intelligence in living beings.
- *Ahamkara* or ego-sense-second product of evolution. It is responsible for the self-sense in living beings.
- *Manas* or instinctive mind-evolves from the *sattva* aspect of *ahamkara*.
- *Panch jnana indriya* or five sense organs-also evolves from the *sattva* aspect of *Ahamkara*.

- *Panch karma indriya* or five organs of action-The organs of action are hands, legs, vocal apparatus, urino-genital organ and anus. They too evolve from the *sattva* aspect of *Ahamkara*
- *Panch tanmatras* or five subtle elements-evolves from the *Tamas* aspect of *Ahamkara*. The subtle elements are the root energies of sound, touch, sight, taste and smell.
- *Panch mahabhuta* or five great substances-ether, air, fire, water and earth. This is the revealed aspect of the physical universe.

The evolution of primal Nature is also considered to be purposeful-Prakrti evolves *for* the spirit in bondage. The spirit who is always free is only a witness to the evolution, even though due to the absence of discriminate knowledge, he misidentifies himself with it. The evolution obeys causality relationships, with primal Nature itself being the material cause of all physical creation. The cause and effect theory of Sankhya is called Satkaarya-vaada (theory of existent causes), and holds that nothing can really be created from or destroyed into nothingness-all evolution is simply the transformation of primal Nature from one form to another.

The evolution of matter occurs when the relative strengths of the attributes changes. The evolution ceases when the spirit realises that it is distinct from primal Nature and thus cannot evolve. This destroys the purpose of evolution, thus stopping Prakrti from evolving for Purusha.

This was a dualistic philosophy. But there are differences between the Samkhya and Western forms of dualism. In the West, the fundamental distinction is between mind and body. In Samkhya, however, it is between the self (purusha) and matter, and the latter incorporates what Westerners would normally refer to as "mind". This means that the Self as the Samkhya understands it is more transcendent than "mind". It is sometimes defined as 'that which observes' and the mind is the instrument through which this observation occurs.

Samkhyan cosmology describes how life emerges in the universe; the relationship between Purusha and Prakriti is crucial to Patanjali's yoga system. The evolution of forms at the basis of Samkhya is quite unique. The strands of Sankhyan thought can be traced back to the Vedic speculation of creation. It is also frequently mentioned in the Mahabharata and Yogavasishta. Sankhya also has a strong cognitive theory built into it; curiously, while consciousness/spirit is considered to be radically different from any physical entities, the mind (*manas*), ego (*ahamkara*) and intellect (*buddhi*) are all considered to be manifestations of Prakrti (physical entity).

There is no philosophical place for a creator God in the Sankhya philosophy; indeed, the concept of God was incorporated into the Sankhya viewpoint only after it became associated with the theistic Yoga system of philosophy. Sankhya serves as the main opponent of Vedanta Philosophy which elucidates the monistic theory of creation. For the Vedanta school, Consciousness/Brahman is the cause of this world. Sankhya denies that vehemently as material world that is insentient cannot originate from a sentient element.

## Yoga

Yoga is a family of ancient spiritual practices that originated in India, where it remains a vibrant living tradition and is seen as a means to enlightenment. Karma Yoga, Bhakti Yoga, Jnana Yoga, and Raja Yoga are considered the four main yogas, but there are many other types. In other parts of the world where yoga is popular, notably the West, Yoga has become associated with the asanas (postures) of Hatha Yoga, which are popular as fitness exercises and also form the basis of an expanding business.

Yoga as a means to enlightenment is central to Hinduism, Buddhism, Jainism and has influenced other religious and spiritual practices throughout the world. Important Hindu texts establishing the basis for yoga include the Upanishads, Yoga Sutras of Patanjali, the Bhagavad Gita, and the Hatha Yoga Pradipika.

***Yoga Practice and Intention:*** In Hinduism, Yoga is considered to be the ultimate way of attaining Enlightenment.

The earliest written accounts of yoga appear in the Rig Veda, which began to be codified between 1500 and 1200 BCE. Some historians believe that this 5000-year-old sculpture is of a yogi.

Modern yoga practice often includes traditional elements inherited from eastern religion, such as moral and ethical principles, postures designed to keep the body fit, spiritual philosophy, instruction by a guru, chanting of mantras (sacred syllables), pranayama (breathing exercises), and stilling the mind through meditation. These elements are sometimes adapted to meet the needs of non-Hindu practitioners, who may be attracted to yoga by its utility as a relaxation technique or as a way to keep fit.

Proponents of yoga see daily practice as beneficial in itself, leading to improved health, emotional well-being, mental clarity, and joy in living. Yoga advocates progress toward the experience of samadhi, an advanced state of meditation where there is absorption in inner ecstasy.

The goals of yoga are expressed differently in different traditions. In theistic Hinduism, yoga may be seen as a set of practices intended to bring people closer to God-to help them achieve union with God. In Buddhism, which does not postulate a creator-type god, yoga may help people deepen their wisdom, compassion, and insight. In Western nations, where there is a strong emphasis on individualism, yoga practice may be an extension of the search for meaning in self, and integration of the different aspects of being. The terms Self-Realization and god-Realization are used interchangeably in Hindu yoga, with the underlying belief that the true nature of self, revealed through the practice of yoga, is of the same nature as God.

The ultimate goal of yoga is the attainment of liberation (Moksha) from worldly suffering and the cycle of birth and death (Samsara). Yoga entails mastery over the body, mind, and emotional self, and transcendence of desire. It is said to lead gradually to knowledge of the true nature of reality. The Yogi reaches an enlightened state where there is a cessation of thought and an experience of blissful union. This union may be of the individual soul (Atman) with the supreme Reality

(Brahman), as in Vedanta philosophy; or with a specific god or goddess, as in theistic forms of Hinduism and some forms of Buddhism. Enlightenment may also be described as extinction of the limited ego, and direct and lasting perception of the non-dual nature of the universe.

For the average person still far from enlightenment, yoga can be a way of increasing one's spiritual awareness, or cultivating compassion and insight. While the history of yoga strongly connects it with Hinduism, proponents claim that yoga is not a religion itself, but contains practical steps which can benefit people of all religions, as well as those who do not consider themselves religious.

***Origins:*** Images of a meditating yogi from the Indus Valley Civilization are thought to be 6 to 7 thousand years old. The earliest written accounts of yoga appear in the Rig Veda, which began to be codified between 1500 and 1200 BC. It is difficult to establish the date of yoga from this as the Rig Veda was orally transmitted for at least a millennium. The first Yoga text dates to around the 2nd century BC by Patanjali, and prescribes adherence to "eight limbs" (the sum of which constitute "Ashtanga Yoga") to quiet one's mind and merge with the infinite.

The first full description of the principles and goals of yoga are found in the Upanisads, thought to have been composed between the eighth and fourth centuries BC. The Upanisads are also called Vedanta since they constitute the end or conclusion of the Vedas (the traditional body of spiritual wisdom). In the Upanisads, the older practises of offering sacrifices and ceremonies to appease external gods gives way instead to a new understanding that man can, by means of an inner sacrifice, become one with the Supreme Being (referred to as Brhman or Mhtman) — through moral culture, restraint and training of the mind.

The word "yoga" derives from the Sanskrit root *yuj* ("to yoke"); which is cognate to modern English "yoke". Both derive from the Proto-Indo-European root yeug-meaning "to join" or "unite".

It is generally translated as "union of the individual atma (loosely translated to mean soul) with Paramatma, the universal soul." This may be understood as union with the Divine by integration of body, mind, and spirit. Thus, in essence, one who attempts yoga may loosely be referred to as a *yogi* or in Sanskrit, a yogin (masculine) or *yogini* (feminine). These designations are actually intended for advanced practitioners, who have already made considerable progress along the path, towards yoga.

***Diversity of Yoga:*** Over the long history of yoga, different schools have emerged, and there are numerous examples of subdivisions and synthesis. It is common to speak of each form of yoga as a "path" to enlightenment. Thus, yoga may include love and devotion (as in Bhakti Yoga), selfless work (as in Karma Yoga), knowledge and discernment (as in Jnana Yoga), or an eight-limbed system of disciplines emphasizing meditation (as in Raja Yoga). These practices occupy a continuum from the religious to the scientific. They need not be mutually exclusive. (A person who follows te path of selfless work might also cultivate some knowledge and devotion.) Some people (particularly in Western cultures) pursue Hatha yoga as exercise divorced from spiritual practice.

Other types of yoga include Mantra Yoga, Kundalini Yoga, Iyengar Yoga, Kriya Yoga, Integral Yoga, Nitya Yoga, Maha Yoga, Purna Yoga, Anahata Yoga, Tantra Yoga, Tibetan Yoga, Sahaja Yoga etc. It is often helpful to check the teacher and lineage to be sure how these terms are being used. Another name for Raja Yoga ("royal yoga") is Ashtanga Yoga ("eight-limbed yoga"), but this should not be confused with the Ashtanga Vinyasa Yoga developed by Sri K. Pattabhi Jois, which is a specific style of Hatha Yoga practice.

***Yoga and Religion:*** In the Hindu, Buddhist and Jain traditions, the spiritual goals of yoga are seen as inseparable from the religions of which yoga forms a part. Some yogis make a subtle distinction between religion and yoga, seeing religion as more concerned with culture, values, beliefs and rituals; and yoga as more concerned with Self-Realization, *i.e.*, direct perception of the ultimate truth. In this sense, religion and yoga are complementary. Sri Ramakrishna likened religion to the husk, and direct experience to the kernel. Both are needed, "but if one wants to get at the kernel itself, he must remove the husk of the grain."

Some forms of yoga come replete with a rich iconography, while others are more austere and minimalist. Hindu practitioners of yoga are proud of their religious traditions, while non-Hindu practitioners claim that yoga may be practiced sincerely by those who have not accepted the Hindu religion.

While the yoga tradition remains rooted in India, the fact that some modern yogis like Swami Vivekananda and Paramahansa Yogananda came to the West suggests that they saw hope the yoga tradition could also flourish there. Critics of yoga as practiced in the West charge that it is sometimes watered down, corrupted, or cut off from its spiritual roots (*e.g.* the popular view that yoga is primarily physical exercises). The sheer number of people practicing yoga outside India suggests the need to define yoga both by its historical roots and its modern adaptations.

***Common Themes:*** Common to most forms of yoga is the practice of concentration (*dharana*) and meditation (*dhyana*). *Dharana*, according to Patanjali's definition, is the "binding of consciousness to a single point." The awareness is concentrated on a fine point of sensation (such as that of the breath entering and leaving the nostrils). Sustained single-pointed concentration gradually leads to meditation (*dhyana*), in which the inner faculties are able to expand and merge with something vast. Meditators sometimes report feelings of peace, joy, and oneness.

The focus of meditation may differ from school to school, *e.g.* meditation on one of the chakras, such as the heart center (anahata) or the 'third eye' (ajna); or meditation on a particular deity, such as Krishna; or on a quality like peace. Non-dualist schools such as Advaita Vedanta may stress meditation on the Supreme with no form or qualities (Nirguna Brahman). This resembles Buddhist meditation on the Void.

Another common element is the spiritual teacher (guru in Sanskrit; lama in Tibetan). While emphasized to varying degrees by all schools of yoga, in some the guru is seen as an embodiment of the Divine. The guru guides the student (*shishya* or *chela*) through yogic discipline from the beginning. Thus, the novice yoga student traditionally devotes himself to a *satguru* (true teacher). Traditionally, knowledge of yoga-as well as permission to practice it or teach it has been passed down through initiatory chains of gurus and their students. This is called *guruparampara*. The yoga tradition is one of practical experience, but also incorporates texts which explain the techniques and philosophy of yoga. Many gurus write on the subject, either providing modern translations and elucidations of classical texts, or explaining how their particular teachings should be followed. A guru may also found an ashram or order of monks; these comprise the institutions of yoga. The yoga tradition has also been a fertile source of inspiration for poetry, music, dance, and art.

When students associate with a particular teacher, school, ashram or order, this naturally creates yoga communities where there are shared practices. Chanting of mantras such as Aum, singing of spiritual songs, and studying sacred texts are all common themes. The importance of any one element may differ from school to school, or student to student. Differences do not always reflect disagreement, but rather a multitude of approaches meant to serve students of differing needs, background and temperament. The yogi is sometimes portrayed as going beyond rules-based morality. This does not mean that a yogi will act in an immoral fashion, but rather that he or she will act with direct knowledge of the supreme Reality· In some legends, a yogi-having amassed merit through spiritual practice, may then cause mischief even to the gods· Some yogis in history have been naked ascetics, such as Swami Trailanga, who greatly vexed the occupying British in 19th century Benares by wandering about naked.

***Roots of Yoga:***

**_Bhagavad Gita_**

**Schools**

Samkhya • Nyaya

Vaisheshika • Yoga

Purva Mimamsa • Vedanta

**Schools of Vedanta**

Advaita • Vishishtadvaita

Dvaita

**Important figures**

Kapila • Gotama

Kanada • Patanjali

Jaimini • Vyasa

Adi Shankara • Ramanuja

Madhva

The Bhagavad Gita famously distinguishes several types of "yoga", corresponding to the duties of different nature of people. Capturing the essence and at the same time going into detail about the various Yogas and their philosophies, it constantly refers to itself as such, the "Scripture of Yoga". The book is thought to have been written some time between the 5th and the 2nd century BC. In it, Krishna describes the following yogas:

(1) Karma yoga, the yoga of "action" in the world.
(2) Jnana yoga, the yoga of knowledge and intellectual endeavor.
(3) Bhakti yoga, the yoga of devotion to a deity.
(4) Raja yoga, the yoga of meditation

***Patanjali:*** Perhaps the classic description of yoga is the Yoga Sutras of Patanjali, which form the basis not only of the darshana called "yoga"—one of six such "orthodox" (*i.e.* Veda-accepting) schools of Hindu philosophy—but also of the practice of yoga in most ashrams (to the extent these can be distinguished). The school (*dharshana*) of Indian philosophy known as "yoga" is primarily Upanishadic with roots in Samkhya, and some scholars see some influence from Buddhism. The Yoga philosophy fully believes in the epistemology of the Samkhya school, as well as its concept of the individual spirits

(Purusha) and the Nature (Prakriti)—but differs from Samkhya's atheism.

Patanjali in the Yoga Sutras presents the goal of yoga as 'the cessation of mental fluctuations' (cittavrtti nirodha), an achievement which gives rise to the possibility of stable meditation and thus deeper states of absorption (dhyana or samadhi). This requires considerable restraint (*yama*) and self-discipline (*niyama*; see below for Patanjali's eight limbs of yoga)). Patanjali's yoga is sometimes called Raja Yoga (Skt: "Royal yoga") or "Ashtanga Yoga" ("Eight-Limbed Yoga"), in order to distinguish it from Hatha yoga. It is held as authoritative by all schools. Patanjali is also known for writing commentaries (Mahabhashya) on the Sutras of the great Sanskrit grammarian Panini. In fact, Panini, Patanjali and Ktyyana are regarded are the highest authority not only in Sanskrit but also in the whole of Linguistics.

Patanjali's text sets forth eight "limbs" of yoga practice. Interestingly, only one of them involves physical postures (and these mainly involve seated positions). The eight are:

(1) Yama (The five "abstentions"): violence, lying, theft, (illicit-) sex, and possessions
(2) Niyama (The five "observances"): purity, contentment, austerities, study, and surrender to god
(3) Asana: This term literally means "seat," and originally referred mainly to seated positions. With the rise of Hatha yoga, it came to be used of these yoga "postures" as well.
(4) Pranayama: Control of *prna* or vital breath
(5) *Pratyahara* ("Abstraction"): Reversal of the sense organs ("That by which the senses do not come into contact with their objects and, as it were, follow the nature of the mind." — Vyasa)
(6) Dharana ("Concentration"): Fixing the attention on a single object
(7) Dhyana ("Meditation")
(8) Samadhi: Super-conscious state or trance (state of liberation)

***God in Yoga Philosophy:*** The philosophy of Yoga also presented certain arguments for the existence of God (Ishvara, lit., the Supreme Lord):

- The Vedas are regarded as evidence. The Vedas and their commentaries, the Upanishads mention and describe god—hence god exists.
- Continuity: people and things have various degrees of differences among themselves. Some people are foolish, some are wise. Hence there ought to be some Being who has the highest level of knowledge among all—who is omniscient. That being is god.
- Cosmic Evolution, leading to this universe, occurs because of the contact between Purusha (spirit) and Prakriti (Nature). Purusha is static, and Prakriti is unconscious. Hence there can be no contact between these two things of opposite characteristics, unless god—the omniscient being—brings about this contact.
- Meditation upon a deity is regarded as the best means of attaining liberation. If meditation on such a being helps in liberation, and all obstacles are removed, then the object of the meditation must have a real existence.

Ishvara is regarded as a special Purusha, who is beyond sorrow and Karma laws. He is one, perfect, infinite, omniscient, omnipresent, omnipotent and eternal. He is beyond the three qualities of Sattva, Rajas and Tamas. He is different from an ordinary liberated spirit, because the latter were bound once, whereas Ishvara was never bound. He is kind and merciful. He is the father of the demigods (the various Devas) and of the sages (rishis), as well as their guru; He is the author of the Vedas. Yoga system is perhaps the first philosophy in the world to give arguments for monotheism. Yoga says that Ishvara can be only one and unique. If many gods are assumed:

- Let us assume there are two gods. If god gives a certain quality (say white colour) to a thing and god gives another (say black colour) to the same thing, this would be mutually contradictory. On the other hand, if god choice reigns supreme, god would fail to remain as god.

- Let us assume the gods work in as a committee to do certain tasks one by one. Then while one god is doing his work, the existence of the other gods would be superfluous and unnecessary.

***Hatha Yoga:*** Over the last century the term yoga has come to be especially associated with the postures (Sanskrit sanas) of hatha yoga ("Forced Yoga"). Hatha yoga has gained wide popularity outside of India and traditional yoga-practicing religions, and the postures are sometimes presented as entirely secular or non-spiritual in nature.

Traditional Hatha Yoga is a complete yogic path, including moral disciplines, physical exercises (*e.g.*, postures and breath control), and meditation-and encompasses far more than the yoga of postures and exercises practiced in the West as a physical culture. The seminal work on Hatha Yoga is the Hatha Yoga Pradipika, written by Swami Svatmarama.

Hatha Yoga was invented to provide a form of physical purification and training that would prepare aspirants for the higher training that is called Raja Yoga (see above). This is still true today. Despite this, many in the West practice 'Hatha yoga' solely for the perceived health benefits it provides, and not as a path to enlightenment.

***Natya Yoga:*** The guide to Natya (Dance) Yoga was written by Bharata Muni. Sage Narada along with Gandharvas were the first to practise Natya Yoga, which comprise all the four main yoga's. Natya Yoga was practised by the medieval devadasis, and is currently taught in a few orthodox schools of Bharatanatyam and Odissi.

***Yoga and Buddhism:*** Within the various schools of Tibetan Buddhism yoga likewise holds a central place, though not in the form presented by Patanjali or the Gita. (For example, physical postures are rarely practiced.) An example would be "guru yoga," the union with the mind of the spiritual teacher which must be done at the beginning of the spiritual path and regularly throughout. In the tantric traditions a number of practices are classified with the name "yoga", for example, the two of the four general classification of tantras-"Yoga Tantra" and "Highest Yoga Tantra".

A system of 108 bodily postures practiced with breath and heart rhythm timing in movement exercises is known as Trul khor or union of moon and sun (channel) prajna energies. The body postures of Tibetan ancient yogis are depicted on the walls of the Dalai Lama's summer temple of Lukhang. As the whole buddhist lineage transmission of Kagyu school came to Tibet over the Indian Yogis Naropa, Tilopa, Marpa then Milarepa, Gampopa, authentic old buddhist yogic practices have been passed over to students still following these instructions throughout many Kagyu Monasteries and institutes worldwide.

Yogacara ("Yoga Adepts"), which is also known as Cittamatra ("Consciousness Only") is an important philosophical school within Indo-Tibetan Buddhism.

***Yoga among Christians:*** Some Christians have changed the practice of yoga to accommodate their own approach to spirituality and out of concern for associating with spiritual practices of other non-Christian religions. Some Christians oppose major components of yoga outright. According to Donal O'Mathuna, Ph.D., and Walt Larimore, M.D., in their book *Alternative Medicine*, they claim: "Yoga is an alternative therapy that is difficult to wholeheartedly accept or reject. As a set of physical and breathing exercises, it can improve general well-being. As a deeply religious practice with the goal of union with the divine, it is antithetical to biblical Christianity."

Other Christians have embraced many aspects of yoga and have incorporated the practice into their ministry, including the Outstretched ministry of Susan Bordenkirche, author of *Yoga For Christians*. Another book is *An Invitation to Christian Yoga* by Nancy Roth.

***Yoga and Tantra:*** Yoga is often mentioned in company with Tantra. While the two have deep similarities, most traditions distinguish them from one another.

They are similar in that both amount to families of spiritual texts, practices, and lineages with origins in the Indian subcontinent. (Coincidentally, both have been popularized to some extent in the West, with perhaps a shallower understanding of their nature). It should be noted however that for the most part, we are speaking of *different* families of texts, lineages, etc.

Their differences are variously expressed. Some Hindu commentators see yoga as a process whereby body consciousness is seen as the root cause of bondage, while tantra views the body as a means to understanding, rather than as an obstruction. It must be said that in India, tantra often carries quite negative connotations involving sexual misbehavior and black magic. Nevertheless, most forms of tantra follow more mainstream social mores. The Hatha Yoga Pradipika is generally classified as a Hindu tantric scripture.

Tantra has roots in the first millennium CE, and incorporates much more of a theistic basis. Almost entirely founded on Shiva and Shakti worship, Hindu tantra visualizes the ultimate Brahman as Param Shiva, manifested through Shiva (the passive, masculine force of Lord Shiva) and Shakti (the active, creative feminine force of his consort, variously known as Ma Kali, Durga, Shakti, Parvati and others). It focuses on the kundalini, a three and a half-coiled 'snake' of spiritual energy at the base of the spine that rises through the chakras until union between Shiva and Shakti (also known as samadhi) is achieved. (Some Hindu yoga teachers, however, have adopted these concepts.)

Tantra emphasises mantra (Sanskrit prayers, often to gods, that are repeated), yantra (complex symbols representing gods in various forms through intricate geometric figures), and rituals that range from simple murti (statue representations of deities) or image worship to meditation on a corpse! While tantric texts (see kaularvatantra, mahanirvana tantra) and teachers (*e.g.* Abhinava Gupta) may seem odd and highly arcane from the point of view of classical yoga, that these incorporate yoga concepts seems clear.

***Notable Yogis:*** Many dedicated individuals have influenced the practice of yoga, and spread awareness of yoga throughout the world. Centuries ago, such individuals included Meera from the Bhakti tradition, Shankaracharya from the Jnana

Yoga tradition, Patanjali, who formalized the system of Raja Yoga, are just a few examples. In the late 1800s, Ramakrishna Paramahamsa, a Bhakti Yogi, brought about a rebirth of yoga in India. A devotee of Mother Kali and a teacher of Advaita Vedanta, he preached that "all religions lead to the same goal."

The noted Indian author Sri Aurobindo translated and interpreted Yogic scriptures, such as the Upanishads and Bhagavad-Gita. His epic poem *Savitri* is a treasure of Hindu Yogic literature, among the longest poems ever written in English. He also founded Sri Aurobindo Ashram in Pondicherry, which continues to propagate the practice of Integral Yoga, which is Aurobindo's synthesis of the four main Yogas (Karma, Jnana, Bhakti and Raja).

Other Indian yogis who inspired their countrymen include Swami Rama Tirtha, and Swami Sivananda who authored over 300 books on yoga and spirituality.

Gopi Krishna was a Kashmiri office worker and spiritual seeker who wrote best-selling autobiographical accounts of his spiritual experiences.

During the early twentieth century, many yogis travelled to the west to spread knowledge of Yoga.

Swami Vivekananda, Ramakrishna's disciple, is well known for introducing Yoga philosophy to many in the west, as well as reinvigorating Hinduism in a modern setting during India's freedom struggle. Swami Sivananda (1887-1963), founder of the Divine Life Society lived most of his life in Rishikesh, India. He wrote an impressive 300 books on various aspects of Yoga, religions, philosophy, spirituality, Hinduism, moral ethics, hygiene and health. He was a pioneering Yogi in bringing Yoga to the west and throughout the world. He was clear, simple and precise in all his teachings. His motto being: "Serve. Love. Give. Meditate. Purify. Realise."

Paramahansa Yogananda (1893-1952), a practitioner of Kriya Yoga, taught Yoga as the binding force that reconciled Hinduism and Christianity. Yogananda founded the Self-Realization Fellowship in Los Angeles, in 1925. His book Autobiography of a Yogi continues to be one of the best-selling books on yoga. A.C. Bhaktivedanta Swami Prabhupada popularised Bhakti Yoga for Krishna in many countries through his movement, the International Society for Krishna Consciousness, (popularly known as the Hare Krishna movement) which he founded in 1966. His followers, known for enthusiastic chanting in public places, brought Bhakti Yoga to the attention of many westerners.

In 1955, the socio-spiritual organization Ananda Marga (the path of bliss) was founded by P.R. Sarkar also known as Shrii Anandamurti. Based on tantric yoga, his teaching emphasizes social service in the context of a political, economic and cultural theory; or "self-realization and service to all."

Also during this period, many yogis brought greater awareness of Hatha yoga to the west. Some of these individuals include students of Sri Tirumalai Krishnamacharya, who taught at Mysore Palace from 1924 until his death in 1989; these students include Sri K. Pattabhi Jois, B.K.S. Iyengar, Indra Devi and Krishnamacharya's son T.K.V. Desikachar.

About the same time, the Beatles' interest in Transcendental Meditation served to make a celebrity of Maharishi Mahesh Yogi.

## Mimamsa

Mimamsa, a Sanskrit word meaning "investigation", is also the name of a school of Indian Philosophy whose primary enquiry is into the nature of dharma. Mimamsa is better known as Purva Mimamsa since it investigates the Purva or "earlier" portions of the Vedas. The foundational text for the Mimamsa school is the Purva Mimamsa Sutras of Jaimini. An important proponent of this school is Kumarila Bhatta. Dharma as understood by Mimamsa can be loosely translated into English as Virtue, Morality or Duty. The Mimamsa school traces the source of the knowledge of dharma neither to sense-experience nor inference, but to verbal cognition (*i.e.* knowledge of words and meanings). Mimamsa believes that all words and meanings exist eternally, and the human mind only perceives them temporarily.

Thus the Rishis or seers *saw* (not with the physical, but the metaphysical eye) the eternal words and meanings of the Vedas that constitute dharma and gave them to mankind as a revelation. The Mimamsa school held dharma to be equivalent to following the commandments of the Vedas and related scriptures such as the Smritis, which involved the performance of Vedic rituals. Seen in this light, Mimamsa is essentially Vedic Ethics, and places great weight on the performance of Karma or action as enjoined by the Vedas.

This school's most valuable contribution to Hinduism was its formulation of well-accepted rules of Vedic interpretation. The beliefs of the Mimamsa school include rejection of a creator God as well as scriptures on dharma outside of the Vedic tradition, and acceptance of the existence of svarga or heaven awaiting the person who has acted righteously in his or her life.

Mimamsa does not pay as much attention to Moksha or salvation as does Vedanta (also known as Uttara Mimamsa-investigation into the Uttara or "later" portions of the Vedas). At a later stage, however, the Mimamsa school changed its views in this regard and began to teach the doctrines of God and Moksha (liberation). While Mimamsa does not receive much scholarly attention these days, its influence can be felt in the life of the practising Hindu. All Hindu ritual, ceremony and religious law is influenced by it.

## Brahman

According to Adi Shankara, God, the Supreme Cosmic Spirit or Brahman is the One, the whole and the only reality. Other than Brahman, everything else, including the universe, material objects and individuals are false. Brahman is at best described as that infinite, omnipresent, omnipotent, incorporeal, impersonal, transcendent reality that is the divine ground of all Being. It (grammatically neutral, but exceptionally treated as masculine), though not a substance, is the basis of the material world, which in turn is its illusionary transformation. Brahman is not the effect of the world. Brahman is said to be the purest knowledge itself, and is illuminant like a source of infinite light. Due to ignorance (*avidy*), the Brahman is visible as the material world and its objects.

The actual Brahman is attributeless and formless. It is the Self-existent, the Absolute and the Imperishable (not generally the object of worship but rather of meditation). Brahman is actually indescribable. It is at best, "Sacchidananda". Also, Brahman is free from any kind of differences. It does not have any *sajtiya* (homogeneous) differences because there is no second Brahman. It does not have any *vijtiya* (heterogeneous) differences because there is nobody in reality existing other than Brahman. It has neither *svagata* (internal) differences, because Brahman is itself homogeneous.

Though Brahman is self-proved, Adi Shankara also proposed some logical proofs:

- *Shruti* — the Upanishads, the Bhagavad Gita and the Brahma Sutras describe Brahman in almost exact manner as Adi Shankara. This is the testimonial proof of Brahman.
- *Psychological* — every person experiences his soul, or atman. According to Adi Shankara, Atman= Brahman. This argument also proves the omniscience of the Brahman.
- *Teleological* — the world appears very well ordered; the reason for this cannot be an unconscious principle. The reason must be due to the Brahman.
- *Essential* — Brahman is the basis of this created world.
- *Perceptible feeling* — many people, when they achieve the *turiya* state, claim that their soul has become one with everything else.

He is the most important contribution of Adi Shankara. *My* is that complex illusionary power of Brahman which causes the Brahman to be seen as the distinct material world. It has two main functions—one is to "cover up" Brahman from the human minds, and the other is to present the material world in its stead. My is also indescribable. It is neither completely real nor completely unreal—hence indescribable. Its shelter is Brahman, but Brahman itself is untouched by the profanity of My, just like a magician is not tricked by his own magic. My is temporary and is destroyed with "true knowledge". This Myvda of Adi Shankara was highly criticized and misunderstood. Bhaskaracharya described Adi Shankara to be indebted to the Buddhists for his concept of My. The term Maya however appears in the Bhagavad Gita 7.14 and many Upanishads.

The concept of My seems to be a hypothesis. Since according to the Upanishads only Brahman is real, but we see the material world to be real, Adi Shankara explained the anomaly by the concept of this illusionary power My.

***Status of the World:*** Advaita teachings that the universe is false often confuse people. Adi Shankara says that the world is not true, it is an illusion, but this is because of some logical reasons. Let us first analyse Adi Shankara's definition of Truth, and hence why the world is not considered true.

- Adi Shankara says that whatever thing remains eternal is true, and whatever is non-eternal is untrue. Since the world is created and destroyed, it is not true.
- Truth is the thing which is unchanging. Since the world is changing, it is not true.
- Whatever is independent of space and time is true, and whatever has space and time in itself is untrue.
- Just as one sees dreams in sleep, he sees a kind of super-dream when he is waking. The world is compared to this conscious dream.
- The world is believed to be a superimposition of the Brahman. Superimposition cannot be true.

On the other hand, Adi Shankara claims that the world is not absolutely false. It appears false only when compared to Brahman. In the pragmatic state, the world is completely true—which occurs as long as we are under the influence of Maya.

The world cannot be both true and false at the same time; hence Adi Shankara has classified the world as indescribable. The following points suggest that according to Adi Shankara, the world is not false (Adi Shankara himself gave most of the arguments):

- If the world were false, then with the liberation of the first human being, the world would have been annihilated. However, the world continues to exist even if a human attains liberation.
- Adi Shankara believes in Karma, or good actions. This is a feature of this world. So the world cannot be false.
- The Supreme Reality Brahman is the basis of this world. The world is like its reflection. Hence the world cannot be totally false.
- False is something which is ascribed to inexistent things, like Sky-lotus. The world is a logical thing which is perceived by our senses.

Consider a scientific logic. A pen is placed in front of a mirror. One can see its reflection. To our eyes, the image of the pen is perceived. Now, what should the image be called? It cannot be true, because it is an image. The truth is the pen. It cannot be false, because it is seen by our eyes.

***Shvara:*** Shvara when man tries to know the attributeless Brahman with his mind, under the influence of Maya, Brahman becomes the Lord. Ishvara is Brahman with Maya— the manifested form of Brahman. Adi Shankara uses a metaphor that when the "reflection" of the Cosmic Spirit falls upon the mirror of Maya, it appears as the Supreme Lord. The Supreme Lord is true only in the pragmatic level — his actual form in the transcendental level is the Cosmic Spirit.

Ishvara is Saguna Brahman or Brahman with innumerable auspicious qualities. He is all-perfect, omniscient, omnipresent, incorporeal, independent, Creator of the world, its ruler and also destroyer. He is causeless, eternal and unchangeable— and is yet the material and the efficient cause of the world. He is both immanent (like whiteness in milk) and transcendent (like a watch-maker independent of a watch). He may be even regarded to have a personality. He is the subject of worship. He is the basis of morality and giver of the fruits of one's Karma. However, he himself is beyond sin and merit. He rules the world with his Maya — his divine power. This association with a "false" knowledge does not affect the perfection of Ishvara, in the same way as a magician is himself not tricked by his magic. However, while Isvara is the Lord of Maya and she (*i.e.*, Maya) is always under his control, the living beings (*jiva*, in the sense of humans) are the servants of Maya (in the form of ignorance).

This ignorance is the cause of the unhappiness and sin in the mortal world. While Ishvara is Infinite Bliss, humans are miserable. Ishvara always knows the unity of the Brahman substance, and the Mayic nature of the world. There is no place for a Satan or devil in Hinduism, unlike Abrahamic religions. Advaitins explain the misery because of ignorance. Ishvara can also be visualized and worshipped in anthropomorphic form as deities such as Vishnu, Krishna or Shiva.

Now the question arises as to why the Supreme Lord created the world. If one assumes that Ishvara creates the world for any incentive, this slanders the wholeness and perfection of Ishvara. For example, if one assumes that Ishvara creates the world for gaining something, it would be against his perfection. If we assume that He creates for compassion, it would be illogical, because the emotion of compassion cannot arise in a blank and void world in the beginning (when only Ishvara existed). So Adi Shankara assumes that Creation is a sport of Ishvara. It is His nature, just as it is man's nature to breathe.

The sole proof for Ishvara that Adi Shankara gives is Shruti's mentions of Ishvara, as Ishvara is beyond logic and thinking. This is similar to Kant 's philosophy about Ishvara in which he says that "faith" is the basis of theism. However, Adi Shankara has also given few other logical proofs for Ishvara, but warning us not to completely rely on them:

- The world is a work, an effect, and so must have real cause. This cause must be Ishvara.
- The world has a wonderful unity, coordination and order, so its creator must have been an intelligent being.
- People do good and sinful work and get its fruits, either in this life or after. People themselves cannot be the giver of their fruits, as no one would give himself the

fruit of his sin. Also, this giver cannot be an unconscious object. So the giver of the fruits of Karma is Ishvara.

***Status of God:*** Some people claim that in Adi Shankara's philosophy, there is no place for a personal God (Ishvara), because Ishvara is also described as "false". He appears as Ishvara because of the curtain of Maya. However, as described earlier, just as the world is true in the pragmatic level, similarly, Ishvara is also pragmatically true. Just as the world is not absolutely false, Ishvara is also not absolutely false. He is the distributor of the fruits of one's Karma. In order to make the pragmatic life successful, it is very important to believe in God and worship him. In the pragmatic level, whenever we talk about Brahman, we are in fact talking about God. God is the highest knowledge theoretically possible in that level. Devotion (Bhakti) will cancel the effects of bad Karma and will make a person closer to the true knowledge by purifying his mind. Slowly, the difference between the worshipper and the worshipped decreases and upon true knowledge, liberation occurs.

***Atman:*** The swan is an important motif in Advaita. It symbolises two things: first, the swan is called *hamsah* in Sanskrit. Upon repeating this *hamso* indefinitely, it becomes *so-aham*, meaning, "I am That". Second, just as a swan lives in water but its feathers are not soiled by water, similarly a liberated Advaitin lives in this world full of maya but is untouched by its illusion.

The soul or the self (Atman) is exactly equal to Brahman. It is not a part of Brahman that ultimately dissolves into Brahman, but the whole Brahman itself. Now the arguers ask that how can the individual soul, which is limited and one in each body, be the same as Brahman? Adi Shankara explains that the soul is not an individual concept. Atman is only one and unique. It is a false concept that there are several Atmans. Adi Shankara says that just as the same moon appears as several moons on its reflections on the surface of water covered with bubbles, the one Atman appears as multiple atmans in our bodies because of Maya. Atman is self-proven, however, some proofs are discussed—*e.g.*, a person says "I am blind", "I am happy", "I am fat" etc. So what is this ego here? Only that thing is the ego which is there in all the states of that person—this proves the existence of Atman, and that consciousness is its characteristic. Reality and Bliss are also its characteristics. By nature, Atman is free and beyond sin and merit. It does not experience happiness or pain. It does not do any Karma. It is incorporeal.

When the reflection of atman falls on Avidya (ignorance), atman becomes *jiva* — a living being with a body and senses. Each jiva feels as if he has his own, unique and distinct Atman, called jivatman. The concept of jiva is true only in the pragmatic level. In the transcendental level, only the one Atman, equal to Brahman, is true.

Adi Shankara exposed the relative nature of the world and established the truth of the Advaita by analysing the three states of experience of the atman — waking (vaishvanara), dreaming (swapna), and deep sleep (sushupti). This idea of a fourth state of consciousness (turiya) apart from these three states is presented in the Mandukya Upanishad.

***Salvation:*** Liberation or Moksha (akin to Nirvana of the Buddhists)—Advaitins also believe in the theory of reincarnation of souls (Atman) into plants, animals and humans according to their karma. They believe that suffering is due to Maya, and only knowledge (called Jnana) of Brahman can destroy Maya. When Maya is removed, there exists ultimately no difference between the Jiva-Atman and the Brahman. Such a state of bliss when achieved while living is called Jivan mukti. While one is in the pragmatic level, one can worship God in any way and in any form, like Krishna as he wishes, Adi Shankara himself was a proponent of devotional worship or Bhakti. But Adi Shankara believes that while Vedic sacrifices, puja and devotional worship can lead one in the direction of jnana, true knowledge, they cannot lead one directly to Moksha.

***Theory of Creation:*** In the pragmatic level, Adi Shankara believes in the Creation of the world through *Satkaryavada*. It is like the philosophy of Samkhya, which says that the cause is always hidden into its effect—and the effect is just a transformation of the cause. However, Samkhya believes in a sub-form of *Satkaryavada* called Parinamvada (evolution)—whereby the cause results in an action. Instead, Adi Shankara believes in a sub-form called *Vivartavada*. According to this, the effect is merely a superimposition of its cause—like its illusion. *e.g.*, In darkness, a man often confuses a rope to be a snake. But this does not mean that the rope has actually transformed into a snake.

At the pragmatic level, the universe is believed to be the creation of the Supreme Lord Ishvara. Maya is the divine magic of Ishvara, with the help of which Ishvara creates the world. The serial of Creation is taken from the Upanishads. First of all, the five subtle elements (ether, air, fire, water and earth) are created from Ishvara. Ether is created by Maya. From ether, air is born. From air, water is born. From water, earth is born. From a proportional combination of all five subtle elements, the five gross elements are created, like the gross sky, the gross fire, etc. From these gross elements, the universe and life are created. This series is exactly the opposite during destruction.

Some people have criticized that these principles are against *Satkaryavada*. According to *Satkaryavada*, the cause is hidden inside the effect. How can Ishvara, whose form is spiritual, be the effect of this material world? Adi Shankara says that just as from a conscious living human, inanimate objects like hair and nails are formed, similarly, the inanimate world is formed from the spiritual Ishvara.

***Status of Ethics:*** Some claim that there is no place for ethics in Advaita, because everything is ultimately illusionary. But on analysis, ethics also has a firm place in this philosophy—the same place as the world and God. Ethics, which implies doing good Karma, indirectly helps in attaining true knowledge. The basis of merit and sin is the Shruti (the Vedas and the Upanishads). Truth, nonviolence, service of others, pity, etc. are Dharma, and lies, violence, cheating, selfishness, greed, etc. are adharma (sin).

***Advaita Vedanta in a Summary:*** Adi Shankara's treatises on the Upanishads, the Bhagavad Gita and the Brahma Sutras are his principal and almost undeniably his own works. Although he mostly adhered to traditional means of commenting on the Brahma Sutra, there are a number of original ideas and arguments. He taught that it was only through direct knowledge of nonduality that one could be enlightened.

Adi Shankara's opponents accused him of teaching Buddhism in the garb of Hinduism, because his non-dualistic ideals were a bit radical to contemporary Hindu philosophy. However, it may be noted that while the Later Buddhists arrived at a changeless, deathless, absolute truth after their insightful understanding of the unreality of samsara, historically Vedantins never liked this idea. Although Advaita also proposes the theory of Maya, explaining the universe as a "trick of a magician", Adi Shankara and his followers see this as a consequence of their basic premise that Brahman is real. Their idea of Maya emerges from their belief in the reality of Brahman, rather than the other way around.

Adi Shankara was a peripatetic orthodox Hindu monk who traveled the length and breadth of India. The more enthusiastic followers of the Advaita tradition claim that he was chiefly responsible for "driving the Buddhists away". Historically the decline of Buddhism in India is known to have taken place long after Adi Shankara or even Kumarila Bhatta (who according to a legend had "driven the Buddhists away" by defeating them in debates), sometime before the Muslim invasion into Afghanistan (earlier Gandhara).

Although today's followers of Advaita believe Adi Shankara argued against Buddhists in person, a historical source, the *Madhaviya Shankara Vijayam*, indicates that Adi Shankara sought debates with Mimamsa, Samkhya, Nyaya, Vaisheshika and Yoga scholars as keenly as with any Buddhists. In fact his arguments against the Buddhists are quite mild in the *Upanishad Bhashyas*, while they border on the acrimonious in the *Brahma Sutra Bhashya*.

The *Vishistadvaita* and *Dvaita* schools believe in an ultimately *saguna* Brahman. They differ passionately with Advaita, and believe that his *nirguna* Brahman is not different from the Buddhist *Sunyata* (wholeness or zeroness) — much to the dismay of the Advaita school. A careful study of the Buddhist Sunyata will show that it is in some ways metaphysically similar as Brahman. Whether Adi Shankara agrees with the Buddhists is not very clear from his commentaries on the Upanishads. His arguments against Buddhism in the Brahma Sutra Bhashyas are more a representation of Vedantic traditional debate with Buddhists than a true representation of his own individual belief.

***The Impact of Advaita:*** Advaita Vedanta philosophy had a tremendous impact on the Hindu system of Tantra and also served to bolster Yogic (see Yoga) ideas of the ultimate Self, Brahman/Atman, being One. Advaita rejuvenated much of Hindu thought and also spurred on debate that led to the expounding of Vishishtadvaita (qualified nondualism) and Dvaita (dualism). Advaita is definitely a very influential philosophy of India.

***Advaita and Science:*** According to some followers of Advaita, it may very well be a place where the scientific world intersects with the spiritual world. They point to the relationships between mass, frequency, and energy that 20th century physics has established and the Advaitic 'Unity of the Universe' as the common ground. They feel that these relationships, formalized as equations by Planck and Einstein, suggest that the whole mesh of the Universe blend into a One that exhibits itself as many (namely, mass, energy, wave etc.), and that this follows Advaita's view that everything is but the manifestation of an omnipresent, omniscient and omnipotent "One".

It must be remembered however, that none of these physicists have talked of an 'omnipresent, omniscient and omnipotent "One"'. They also connect the De Broglie waves of modern physics to Aum in Hindu philosophy. However, scientists in India and abroad clarify that the de Broglie waves (or matter waves) are neither optical nor acoustic waves, but are "just functions of a probability distribution of finding a particle, which may be represented as a Fourier sum of constituent probability waves."

However, notable scientists like Erwin Schrodinger and Robert Oppenheimer were also Vedantists. Fritjof Capra's book, The Tao of Physics, is one among several that pursue this viewpoint as it investigates the relationship between modern, particularly quantum physics and the core philosophies of various Eastern religions, including Hinduism, Buddhism and Taoism.

It must be noted that Advaita does not share the same ground on science as other schools of philosophy do. For example, Adi Shankara rejected the idea of momentariness of the universe in his Brahma Sutra commentary since Brahman is immanent in the Universe, while Buddhists affirm that the universe on its own accord, due to the causality of the dharmas, is constantly changing. The Dvaita-enthusiasts on the contrary, blame Adi Shankara for inconsistency, since he adopts the view that the Universe is momentary in many of his other works like the *Upanishad Bhashya*. Dvaita-enthusiasts see the Universe as a creation of God, while Advaitins see it as a manifestation of Brahman; Buddhists on the other hand see it as a flux of changes, originating from natural phenomena leading to its formation.

***Mahavakya:*** Mahavakya, or "the great sentences", state the unity of Brahman and Atman. They are four in number and their variations are found in other Upanishads.

***Visistadvaita:*** Vishisthadvaita is a qualified monism in which God alone exists but admits plurality. By holding such beliefs, Vishisthadvaita is said to be the midway between Advaita and Dvaita by few. The book, "All about Hinduism" by Swami Sivananda discusses aspects of Vishisthadvaita, a theological school whose most important proponent was Ramanuja.

***The Philosophy:*** In more specific terms Vishishtadvaita conclusions may be briefly summarised as below.

The Absolute Supreme Reality referred to as Brahman, is a Transcendent Personality with infinite superlative qualities. He is Lord Vishnu, also known as Narayana. He creates the other two members of the Trimurti, namely, Creator Brahma and Shiva, the Lord of Deluge. Narayana is the Absolute God. The Soul and the Universe are only parts of this Absolute and hence, Vishishtadvaita is panentheistic. The relationship of God to the Soul and the Universe is like the relationship of the Soul of Man to the body of Man. Individual souls are only parts of Brahman. God, Soul and Universe together form an inseparable unity which is one and has no second. This is the non-duality part.

Matter and Souls inhere in that Ultimate Reality as attributes to a substance. This is the qualification part of the non-duality.

Souls and Matter are only the body of God. Creation is a real act of God. It is the expansion of intelligence. Matter is fundamentally real and undergoes real revelation. The Soul is a higher mode than Matter, because it is conscious. It is also eternally real and eternally distinct. Final release, that comes, by the Lord's Grace, after the death of the body is a Communion with God. This philosophy believes in liberation through one's Karmas (actions) in accordiance with the Vedas, the Varna (caste or class) system and the four Ashramas (stages of life), along with intense devotion to Vishnu. Individual Souls retain their separate identities even after moksha. They live in Fellowship with God either serving Him or meditating on Him. The philosophy of this school is Sri Vaishnavism, a branch of Vaishnavism.

The succession of great Master-Expositors and spiritual giants of Vishishtadvaita school starts with the twelve Alwars, who left behind an imperishable legacy of Tamil devotional poetry in the form of 4000 songs, now called the Nalayira-prabandham.

***Natha Muni:*** Natha-muni of the ninth century AD, the foremost Acharya of the Vaishnavas, collected the Tamil prabandhams, classified them, made the redaction, set the hymns to music and spread them everywhere. He is said to have received the divine hymns straight from Nammalvar, the foremost of the twelve Alwars, by yogic insight in the temple at Alwar Thirunagari, which is located near Tirunelveli.

***Yamunacharya:*** Yamunacharya, who renounced a kingship and spent his last days in the service of the Lord at Srirangam and in laying the fundamentals of the Vishishtadvaita philosophy by writing four basic works on the subject.

***Ramanuja:*** Ramanuja was the first propagator of Vishishtadvaita philosophy on a national scale. He undertook three major tasks of his Master upon himself. The first was his masterly commentary called the Shri-Bhashya, on the Brahma Sutras where he claimed to have proven that the interpretations of the Upanishads according to the Advaita school were not tenable. His second vow was to perpetuate the name of Parashara, the author of Vishnu purana. This he did by calling Bhatta, the son of one of his disciples by that name and making him write a commentary on Vishnu Sahaśranama.

The third vow was to perpetuate the name of Nammalwar, which he did by causing a commentary to be written on Nammalwar's Tamil Veda, Tiruvai-mozhi, which represents the cream of Vishishtadvaita philosophy. The message that Ramanuja left for posterity and the world went home not only in his homeland, but in the entire country up to distant Kashmir. Sri Ramanuja spent about 14 years of the most productive part of his life in Melukote near Mysore, Karnataka, where he could continue his work on Vishishtadvaita in peace under the patronage of the maharajahs of Mysore.

***Vedanta Desika:*** Vedanta Desika, one of the foremost learned scholars of medieval India, wrote more than a hundred works in Sanskrit and Tamil. All are characterised by their versatility, deep spiritual insight, ethical fervour and excellent expressions of devotional emotion in delightful style. His Paduka-sahasram is a classic example. He was a great teacher, expositor, debater, poet, philosopher, thinker and defender of the faith of Vaishnavism.

***Pillai Lokacharya:*** Pillai Lokacharya, who is considered as the real founder of the tenkalai (Southern Learning) sect of Vaishnavism.

***Dvaita:*** Dvaita (originally called Tattvavada), a school of Vedanta (the most widespread Hindu philosophy) founded by

Shri Madhvacharya, stresses a strict distinction between God and souls. According to Madhva, souls are not created by God but depend on Him to evolve or transform. Satguru Sivaya Subramuniyaswami explains dualism in his book, *Dancing with* Siva, by stating that in dualism, God is seen as the efficient cause of the universe and not as the material cause. He is the potter causing the clay to emerge, rather than serving as the source of the clay itself.

***Dvaita Philosophy:*** Like Ramanuja Shri Madhvacharya espoused a Vaishnava theology that understands Brahman to be endowed with attributes and a personal God, Vishnu. By Brahman, he referred to Vishnu, as per his statement (brahmashabdashcha vishhnaveva) that Brahman can only refer to Vishnu. Madhva states that Vishnu is not just any other deity, but is rather the singular, all-important and supreme one. Vishnu is always the primary object of worship, and all others are regarded as subordinate to Him. The deities and other sentient beings are graded among themselves, with Vayu, the god of life, being the highest, and Vishnu is eternally above them.

Dvaita, or Dualistic philosophy, also known as Bheda-vada, Tattva-vada, and Bimba-pratibimba-vada, is the doctrine propounded by Ananda Tirtha (also known as Madhvacharya) that asserts that the difference between the individual soul or jiva, and God, (Ishvara or Vishnu), is eternal and real. Actually, this is just one of the five differences that are so stated — all five differences that constitute the universe are eternal.

The Dvaita doctrine was summarized by Vyasa Tirtha as comprising nine tenets or prameyas.

### *Impact of Dvaita Movement*

- Madhva's Dualistic view, along with Shankara's Advaita or Nondualism and Ramanuja's Qualified Nondualism or Vishishtadvaita form some core Indian beliefs on the nature of reality.
- Madhva is considered to be one of the influential theologians in Hindu history. He revitalized an Hindu monotheism in light of attacks, theological and physical,

by foreign invasion. Great leaders of the Vaishnava Bhakti movement, in Karnataka, for example, Purandara Dasa and Kanaka Dasa were strong proponents of the Dvaita tradition. Also, the famous Hindu saint, Raghavendra Swami, was a leading figure in the Dvaita tradition.

- Madhva's theology heavily influenced those of later scholars such as Nimbarka, Vallabha, and Caitanya. B.N.K. Sharma notes that Nimbarka's theology is a loose re-write of Madhva's in its most essential aspects. Vallabha even "borrowed without acknowledgement" a verse from Madhva's sarva-shastrartha-sangraha'. The followers of Caitanya claim a link to Madhva, though such a link is not historically tenable or theologically plausible.
- Madhva's singular contribution was to offer a new insight and analysis of the classical Vedantic texts — the Vedas, Upanishads, Brahma Sutra, Mahabharata, Pancharatra, and Puranas — and place uncompromising duality, which had been ravaged by attacks from Advaita, on a firm footing. Before Madhva, Nondualism was rejected by others such as the Mimamsa tradition of Vedic exegesis, and by the Nyaya tradition of classical logic. However, it was only he who could build a cogent alternative system of Vedanta that could take on Advaita in full measure.

***Comparison to Mainstream Hinduism:*** Some teachings of Sri Madhvacharya look different from mainstream Hinduism. One example is his doctrine of eternal damnation. Generally Hindus believe in the eventual salvation of every soul. Many of the doctrines in the Dvaita traditions resemble those of strict monotheism that is predominant amongst followers of Semitic religions. Dvaita provides a greater role to Bhakti than other schools of Vedanta. Followers of Dvaita believe in the supremacy of Lord Vishnu over other deities including Shiva and do not believe in the Hindu concept of Trinity, Trimurti of Brahma-Vishnu-Shiva. Shiva is worshipped as a subordinate god. Though this appears intolerant, it is because of the strong monotheistic belief in a non-Impersonal God unlike Advaita for which the identity of God does not matter as it is Nirguna. Historically, Dvaita scholars have been involved in vigorous debates against other schools of thought, especially Advaita. Whereas Adviata preaches that Atman and Brahman are one and the same, which is not evident to the atman till he comes out of a so-called illusion, Sri Madhvacharya puts forward the truth as Brahman (Sri Vishnu) and Atman (soul) to be eternally different, with God always the most superior one.

It is the same point that Sri Madhvacharya reinforces in one of his doctrines... "Yadi Namaparo Na bhaveth Sri Hari, khathamasya vasheth Jagathedhabhooth. Yadi Namanathasya Vashe Sakalam, Khatamevathu nithya suknham Na Bhaveth"

The translation of the above is :

*"If you feel there is no God,how do you explain as to why you cannot free yourself from the limitations on Earth? If you feel you are the one in control of everything (as Advaita preaches that Soul and God are one and the same), then how come you don't enjoy happiness always and are also subject to sorrow and pain (as God is supposed to be an eternity of happiness)?".*

***Maharshi Kapil:*** Maharishi Kapila is a Vedic sage traditionally considered to be the original proponent of the *Samkhya* system of philosophy but there are no known writings by him that deal with the *Samkhya* system as it is understood today. He is described as an incarnation of Vishnu within the Puranas, famous for teaching a process of liberation known as bhakti yoga.

***History:*** Very little historical data is known for sure regarding Maharishi Kapila's life. He is said to have lived in the Indian subcontinent, some say around 500 BC, other accounts give much earlier dates. Tradition has it that Shakyamuni Buddha studied the Samkhya system before his "awakening," putting Kapila's birth at least before that time. He is also mentioned by Krishna in the Bhagavad Gita as the greatest of all perfected beings which could possibly move the date back further still:

*Of all trees I am the banyan tree, and of the sages among the demigods I am Narada. Of the Gandharvas I am Citraratha, and among perfected beings I am the sage Kapila.* *(Bhagavad Gita 10.26)*

His parents are given in the Bhagavata Purana as Kardama Muni and Devahuti. After his father left home, Kapila instructed his mother Devahuti in the philosophy of yoga and worship of Lord Vishnu, enabling her to achieve both liberation (moksha), and pure love of God.

***Teachings:*** Kapiladev's teachings are quoted exstensiveley within the Srimad Bhagavatam especially:

- *"My appearance in this world is especially to explain the philosophy of Sankhya, which is highly esteemed for self-realization by those desiring freedom from the entanglement of unnecessary material desires. This path of self-realization, which is difficult to understand, has now been lost in the course of time. Please know that I have assumed this body of Kapila to introduce and explain this philosophy to human society again."*
- *"When one is completely cleansed of the impurities of lust and greed produced from the false identification of the body as "I" and bodily possessions as "mine," one's mind becomes purified. In that pure state he transcends the stage of so-called material happiness and distress." (3.25.16)*
- *"The Supreme Personality of Godhead is the Supreme Soul, and He has no beginning. He is transcendental to the material modes of nature and beyond the existence of this material world. He is perceivable everywhere because He is self-effulgent, and by His self-effulgent luster the entire creation is maintained."*
- *"The glory of the Lord is always worth singing, for His glories enhance the glories of His devotees. One should therefore meditate upon the Supreme Personality of Godhead and upon His devotees. One should meditate on the eternal form of the Lord until the mind becomes fixed."*

***Birth of the Ganges:*** Maharshi Kapila is a major figure in the story associated with the Hindu holiday of Makar Sankranti, about the birth of the Ganges River, which involves King Sagar of Ayodhya, an ancestor of Rama. King Sagar had performed the Aswamedha sacrifice ninety-nine times. Each time a horse was sent around the earth Indra the King of the Heaven grew jealous and kidnapped the horse, hiding it in the hermitage of Kapila Muni during the hundredth sacrifice.

The 60,000 sons of Sagar found the horse, and believing Kapila to be the abductor assaulted him. Kapila turned the assailants to ashes. Bhagirath, a grandson of King Sagar, came to Kapila begging him to redeem the souls of the 60,000. Kapila replied that only if the Ganges descended from heaven and touched the ashes of the 60,000 would they be redeemed.

***Gotama:*** Gotama was an Indian philosopher and logician who wrote the Nyya Sutras, considered to be the foundation of the Nyya school of philosophy.

Gotama's precise dates are uncertain. In the past he has been assigned to various times between the sixth and third centuries BCE, though modern scholarship places him between the second and third centuries CE. His name has occasionally led him to be confused with Gautama Buddha. He is sometimes given the honorific titles "Akapda" ("eye-footed") and "Dirghatapas" ("long-penance"). He is also sometimes accorded the religious titles "Rishi" or "Maharshi".

In the *Nyya Sutras* Gotama developed and extended the Vaiaeika epistemological and metaphysical system through 528 aphorisms. Later commentaries expanded, expounded, and critically discussed Gotama's work, the first being by Vtsyyana (c.450–500 CE), followed by the *Nyyavrttika* of Uddyotakra (c. 6th–7th century), Vcaspati Mira's *Ttparyatik* (9th century), Udayana's *Ttparyapariuddhi* (10th century), and Jayanta's *Nyyamanjari* (10th century). These not only discussed Gotama's thought, but also that of his successors.

***Kanada:*** Kanada was a Hindu sage who founded the philosophical school of Vaisheshika. He is considered as the father of Atom theory. He probably lived around 600 BCE according to some accounts. It is believed that he was born in Prabhas Kshetra (near Dwaraka) in Gujarat, India. His area of study was Rasavtam, considered to be a type of alchemy. In his text, the Vaishashik Darshana, he describes an atomic theory more than a century before Democritus developed one in ancient Greece. He believed that all objects were comprised of nine elements: earth, water, light, wind, ether, time, space, mind and soul

***Patanjali:*** Patanjali is the compiler of the *Yoga Sutra*, a major work containing aphorisms on the practical and philosophical wisdom regarding practice of Raja Yoga. ("Yoga" in traditional Hinduism involves inner contemplation, a rigorous system of meditation practice, ethics, metaphysics, and devotion to the one common soul, God, or Brahma/Atman.)

Patanjali is the reputed author of the *Yoga Sutra*, as well as commentary on Pnini's Sanskrit grammar which is called 'Mahabhashya'. There are also many ayurvedic texts attributed to him. Nearly all scholars believe these texts were actually written by different people in different eras.

Desiring to teach yoga to the world, he is said to have fallen (pat) from heaven into the open palms (anjali) of a woman, hence the name Patanjali.

The *Yoga Sutra* probably dates around BC 200. Patanjali has often been called the founder of Yoga because of this work, although in reality is more a major figure. The *Yoga Sutra* is a treatise on Raja Yoga, built on the Samkhya school and the Hindu scripture of the Bhagavad Gita. Yoga is also found in the Puranas, Vedas and Upanishads. Still, this work is certainly a major work among the great Hindu scriptures and certainly is the basis of Raja Yoga. Patanjali's Yoga is one of the six schools of Hindu Philosophy. They give us the earliest reference to the popular term Ashtanga Yoga which translates literally as the eight limbs of yoga. They are yama, niyama, asana, pranayama, pratyahara, dharana, dhyana and samadhi.

***The Mahabhashya:*** The Mahabhashya (great commentary) of Patanjali on the celebrated Ashtadhyayi of Panini is one of the three most famous works in Sankrit grammar. It was with Patanjali that Indian linguistic science reached its definite form. The system thus established is extemely detailed as to shiksha (phonology, including accent) and vyakarana (morphology). Syntax is scarecely touched, but nirukta (etymology) is discussed, and these etymologies naturally lead to semantic explanations. People interpret his work to be a defense of Panini, whose Sutras are elaborated meaningfully. He also attacks Ktyyana rather severely. But the main contributions of Patanjali lies in the treatment of the principles of grammar enunciated by him.

***Jamini:*** Maharshi Jaimini was a student of Vyasa Maharishi, whom Jaimini refers to by the name "Badarayana". Jaimini is believed to have lived around the 32nd century BC, at the beginning of the period known, in the Hindu calendar, as the Kali Yuga. His sutras include not only the Rudramsa, Shasthamsa and Saptamsa, but also lagnas (like Varnada Lagna) and yogas (such as Shakti Yoga and Raja Yoga). His Purva Mimamsa Sutras contains about 3,000 sutras discussing duty (dharma), and is the foundational text of the Mimamsa school. Though there had been earlier efforts to systematize knowledge by teachers such as Alekhana and Karsnajini (mentioned by Jaimini), their works have not been passed on. Jaimini's definition of dharma is an act which is endorsed by the Vedas and related scriptures and conducive to human well-being.

***Vyasa:*** Vyasa is an important and much revered figure in the Hindu tradition and its literature. He is considered to be an ideal Brahmarishi-omniscient, truthful, purest of the pure and possessor of knowledge of the essence of Brahman.

***The Legend of Vyasa:*** Vyasa appears for the first time as the author and an important character in the Hindu epic Mahabharata. The epic, whilst having its roots in historical events centuries before the common era, is an extraordinarily long compendium of legend, myth, philosophy and semi-historical material about ancient India. It is impossible to point out when the 'Historical' Vyasa lived or disentangle a factual story from the legendary material contained in the epic. According to the Mahabharata, he was the son of Satyavati, a ferryman's daughter, and the wandering sage Parashara. He was born on an island in the River Yamuna. The place is near a town Kalpi in Jaluan district in Uttar Pradesh. He was dark in colour and hence was named Krishna; as he was born on an island (Sanskrit Dweepa), he was also called Dwaipayana. Thus he is generally known as Krishna-Dwaipayana. The child grew up to be an adult as soon as he was born; adopting the life of an ascetic, he soon became one of the greatest Rishi (Sages).

In the Puranas, especially the Vishnu-Purana and Bhagavata-Purana, he is considered an incarnation—Avatara of Vishnu.

Vyasa is considered to be one of the seven immortals in Hindu mythology.

***'Veda' Vyasa:*** It is traditionally held by Hindus that the sage classified the primordial single Veda into four. Hence he was called Veda Vyasa, or "Splitter of the Vedas," the splitting being a feat that allowed mortals to understand the divine knowledge of the Veda. The word vyasa means split, differentiate, or describe. This title is the most popular way of referring to him.

It has been debated whether Vyasa was a single person or a class of scholars who did the splitting. The Vishnu-Purana has an interesting theory about Vyasa. The Hindu view of the universe is that of a cyclic phenomenon that comes into existence and dissolves repeatedly. Each cycle is presided over by a number of Manus, one for each Manvantara, that has four ages, Yugas of declining virtues. The Dvapara is the third Yuga. The Purana says:

> *In every third world age (Dvapara), Vishnu, in the person of Vyasa, in order to promote the good of mankind, divides the Veda, which is properly but one, into many portions. Observing the limited perseverance, energy, and application of mortals, he makes the Veda four-fold, to adapt it to their capacities; and the bodily form which he assumes, in order to effect that classification, is known by the name of Veda-vyasa. Of the different*

*Vyasas in the present Manvantara and the branches which they have taught, you shall have an account. Twenty-eight times have the Vedas been arranged by the great Rishis in the Vaivasvata Manvantara... and consequently eight and twenty Vyasas have passed away; by whom, in the respective periods, the Veda has been divided into four. The first... distribution was made by Svayambhu (Brahma) himself; in the second, the arranger of the Veda (Vyasa) was Prajapati... (and so on up to twenty-eight).*

***Author of Mahabharata:*** Vyasa is traditionally known as author of this epic. But he also features as an important character in it. His mother later married the king of Hastinapura, and had two sons. Both sons died without an issue and taking recourse to an ancient practice called Niyoga where a chosen man can father sons with the widow of a person who dies issueless, she requests Vyasa to produce sons on behalf of her dead son Vichitravirya. Vyasa fathers the princes Dhritarashtra and Pandu (by Ambika and Ambalika, the wives of dead king. The sequence of events also leads to a third son, Vidura, by a serving maid to the queens. While these are 'legally' not his sons, another son Shuka, born of a celestial nymph, is considered his true spiritual heir. He thus was the grandfather of both the warring parties of the Mahabharata, the Kauravas and the Pandavas. He makes occasional appearances in the story as a spiritual guide to the young princes.

In the first book the Mahabharata, and is said that he asked Ganesha to aid him in writing it down. It is said that Ganesha imposed a condition that Vyasa narrate the story without pause, and Vyasa made a counter-condition that Lord Ganesh understand the verse before he transcribed it. This is supposed to explain the complicated Sanskrit used in some sections of the Mahabharata, recited by Vyasa when he wanted a break.

***Author of Puranas:*** In addition to the epic, he is also credited with the writing of the eighteen major, if not all, Puranas. His son Shuka is the narrator of the major Purana Bagavata-Purana.

***Vyasa in Buddhism:*** Vyasa appears as Kanha-dipayana (the Pali version of his name) in two Jataka tales: the Kanha-dipayana Jataka and Ghata Jataka. Whilst the former in which he appers as the Bodisattva has no relation to his tales from the Hindu works, his role in the latter one has parallels in an important event in the Mahabhrata.

In the 16th book of the epic, Mausala Parva, the end of the Vrishnis, clansmen of Vyasa's namesake and Vishnu incarnate Krishna is narrated. The epic says:

*One day, the Vrishni heroes saw Vishvamitra, Kanwaand Narada arrived at Dwaraka. Afflicted by the rod of chastisement wielded by the deities, those heroes, causing Samba to be disguised like a woman, approached those ascetics and said, 'This one is the wife of Vabhru of immeasurable energy who is desirous of having a son. Ye Rishis, do you know for certain what this one will bring forth?Those ascetics, attempted to be thus deceived, said: 'This heir of Vasudeva, by name Samba, will bring forth a fierce iron bolt for the destruction of the Vrishnis and the Andhakas.*

The important Bhagavata-Purana narrates:

*The sages Visvamitra, Asita, Kanva, Durvasa, Bhrigu, Angira, Kasyapa, Vamadeva, Atri, Vasishthha, along with Narada and others, [once] stayed in the house of the lord of the Yadus [Krishna]...The young boys of the Yadu dynasty playing [there] approached them with Samba the son of Jambavati dressed up in woman's clothes. Taking hold of their feet they, feigning humility, impudently asked: 'This black-eyed pregnant woman wishing for a son, o learned ones, too embarrassed to ask it herself, is asking you whether you, with your vision never clouded, can tell if she'll give birth to a son or not?' The sages thus tricked said angered to the boys, o Kng: 'For you, o fools, she'll give birth to a mace which will destroy the dynasty!*

The Ghata Jataka has a different spin on it:

*The Vrishnis, wishing to test Kanha-dipayana's powers of clairvoyance, played a practical joke on him. They tied a pillow to the belly of a young lad, and dressing him up as a woman, took him to the ascetic and asked when the baby would be born. The ascetic replied that on the seventh day the person before him would give birth to a knot of acacia wood which would destroy the race of Vasudeva. The youths thereupon fell on him and killed him, but his prophecy came true.Notably, he is not the Bodhisattva in the Ghata Jataka.*

***In the Arthashastra:*** The only non-religious book in which Vyasa has an interesting entry is the Arthashastra of Chanakya. In chapter 6, it says:

*Whosoever is of reverse character, whoever has not his organs of sense under his control, will soon perish, though possessed of the whole earth bounded by the four quarters. For example: Bhoja, known also by the name, Dandakya, making a lascivious attempt on a Brahman maiden, perished along with his kingdom and relations; so also Karala, the Vaideha... Vátapi in his attempt under the influence of overjoy to attack Agastya, as well as the corporation of the Vrishnis in their attempt against Dvaipayana.'*

This reference matches the Jataka version and has lead some scholars to speculate that in the older versions Vyasa died in this attack, though the Arthashastra is actually speaking about the destruction of the attackers.

***Author of Brahma Sutra:*** The Brahma-Sutra is attributed to Badarayana, which makes him the proponent of the crest-jewel school of Hindu philosophy, *i.e.*, Vedanta. As the island on which Vyasa was born is said to have been covered by Badara (Indian jujube) trees, he is known as Badarayana. Though traditionally, Vyasa is considered the Badarayana who wrote the Sutras, many historians think these were two different personalities.

***Author of Yoga Bhashya:*** This text is a commentary on the Yoga Sutras of Patanjali. Vyasa is credited with this work also, though this is impossible, if Vyasa's immortality is not considered, as it is a later text.

***Adi Sankara:*** Adi Shankara also known as di Zaekarcrya (the first Shankara in his lineage) and Zaekara Bhagavatpdcrya (the teacher at the feet of God), c. 788 – 820 CE, was the first philosopher in the tradition of Advaita Vedanta, a sub-school of the Vedanta school of Hindu philosophy, to formulate its doctrine. He taught the unity of the soul and God. He viewed God as simultaneously personal and attributeless. He held seminal debates with other philosophers to refute their doctrines. In the Smrta tradition, he is regarded as an incarnation of the god, Shiva.

Born in the South Indian state of Kerala, he toured India, propagating his teachings. He founded four mathas (abbeys) at the four cardinal directions. These institutions, mainly followed by Hindu Smartas, play an important role in the propagation of the Advaita Vedanta philosophy and Hinduism.

***Life:*** The traditional source for accounts of Adi Shankara's life are the Shankara Vijayams (IAST *Zaekara Vijayas*). Literally meaning *Victory of Shankara*, these poetical works contain much historical material interspersed with legends. The most important among these biographies are the *Mdhaviya Zaekara Vijayac-*, the *nandagiriya Zaekara Vijayac*, the *Cidvilsiya Zaekara Vijayac-*, and the *Keraiya Zaekara Vijayac-*. According to these texts, Adi Shankara was born in Kalady, Kerala, India, to a Namboothiri brahmin couple, Shivaguru (IAST *Zivaguru*) and Aryamba (IAST *rymb*) and that he lived for thirty two years.

***Birth and Childhood:*** Adi Shankara's parents were childless for many years. They prayed at the Vadakkumnathan (Vaccala) temple in Thrissur, Kerala, for the birth of a child. Legend has it that Shiva appeared to both husband and wife in their dreams, and offered them a choice: a mediocre son who would live a long life, or an extraordinary son who would not live long. Both the parents chose the latter; thus a son was born to them. He was named Shankara (bestower of happiness), in honour of Shiva (one of whose epithets is Shankara).

His father died while Shankara was very young. Shankara's upanayan, the inititiation into student-life, was performed at the age of five. As a child, Shankara showed remarkable scholarship, mastering the four Vedas by the age of eight. Following the customs of those days, Shankara studied and lived at the home of his teacher. It was customary for students and men of learning to receive Bhikc (alms) from the laity; on one occasion, while accepting Bhikc, Shankara came upon a woman who had only a single dried amalaka fruit to eat. Rather than consuming this last bit of food herself, the pious lady gave away the fruit to Shankara as Bhikc. Moved by her piety, Shankara composed the Kanakadhr Stotram on the spot. Legend has it that on completion of the stotra, golden amalaka fruits were showered upon the woman by Lakcmi, the Goddess of wealth.

***Sannyasa***

***Hindu philosophy***

**Schools**

Samkhya • Nyaya

Vaisheshika • Yoga

Purva Mimamsa • Vedanta

**Schools of Vedanta**

Advaita • Vishishtadvaita

**Dvaita**

**Important figures**

Kapila • Gotama

Kanada • Patanjali

Jaimini • Vyasa

Adi Shankara • Ramanuja

Madhva

From a young age, Shankara was attracted to sannyasa (monastic life). His mother was against his becoming an monk, and refused him her formal permission. However an incident occurred that resulted in her giving assent to his monkhood. Shankara was bathing in the Purna River near his house, when a crocodile gripped his leg and began to drag him into the water. Only his mother was nearby, and it proved impossible for her to rescue him. Shankara then told his mother that he was on th verge of death; if she would give him her formal permission verbally, he would at this moment renounce the world and die a monk. At the end of her wits, his mother agreed; Shankara immediately recited the mantras that made a renunciate of him and awaited death. However, the crocodile released him and swam away. Shankara emerged unscathed from the river.

With the permission of his mother, Shankara left Kerala and travelled towards North India in search of a Guru. On the banks of the Narmada River, he met Govinda Bhagavatpada, the disciple of Gaudapada. When Govinda Bhagavatpada asked Shankara's identity, he replied with an extempore verse that brought out the Advaita Vedanta philosophy. Govinda Bhagavatapada was impressed and took Shankara as his disciple.

Adi Shankara was commissioned by his Guru to write a commentary on the Brahma Sutras and propagate the Advaita Vedanta philosophy. The Madhaviya Shankaravijaya states that Adi Shankara calmed a flood from the Reva River by placing his *kamanalu* (water pot) in the path of the raging water, thus saving his Guru, Govinda Bhagavatpada, who was engaged in Samdhi (meditation) in a cave nearby. On his mission to spread the Advaita Vedanta philosophy, Adi Shankara travelled to Kashi, where a young man named Sanandana from Choladesha in South India, became his first disciple. In Kashi, Adi Shankara had an encounter with an untouchable-on his way to the Vishwanath Temple, he came upon an untouchable with four dogs. When asked to move aside by Shankara's disciples, the untouchable replied: "Do you wish that I move my ever lasting tman (Self), or this body made of food?" Understanding that the untouchable was none other than god Shiva, and his dogs the four Vedas, Shankara prostrated himself before him, composing five shlokas known as Manisha Panchakam.

On reaching Badari in the Himalayas, he wrote the famous *Bhashyas* (commentaries) and *Prakarana granthas* (philosophical treatises). Afterwards he taught these commentaries to his disciples. Some, like Sanandana, were quick to grasp the essence; the other disciples thus became jealous of Sanandana. In order to convince the others of Sanandana's inherent superiority, Adi Shankara summoned Sanandana from one bank of the Ganga River, while he was on the opposite bank. Sanandana crossed the river by walking on the lotuses that were brought out wherever he placed his foot. Adi Shankara was greatly impressed by his disciple and gave him the name Padmapda (lotus-footed one). The sage, Vedavysa, visited Adi Shankara in the guise of an old Brhmana. Adi Shankara debated with the Brhmana for over eight days when at last, Vyasa revealed his real identity and blessed Adi Shankara.

***Meeting with Mandana Mishra:*** One of the most famous debates of Adi Shankara was with the ritualist Mandana Mishra (IAST *Manana Mira)*. Mandana Mishra's Guru was the famous Mimamsa philosopher, Kumarila Bhamma. Shankara sought a debate with Kumarila Bhamma and met him in Prayag where he had buried himself in a slow burning pyre to repent for sins committed against his Guru: Kumarila Bhamma had learnt Buddhist philosophy incognito from him in order to be able to refute it. This constitues a sin according to the Vedas. Kumarila Bhamma thus asked Adi Shankara to proceed to Mahicmati (known today as Maheshwar in Madhya Pradesh) to meet Mandana Mishra and debate with him instead.

Adi Shankara had a famous debate with Mandana Mishra in which the wife of Mandana Mishra, Ubhaya Bhrati, was the referee. After debating for over fifteen days, Mandana Mishra accepted defeat. Ubhaya Bhrat+ then challenged Adi Shankara to have a debate with her in order to 'complete' the victory. This debate was to be on the subject of *kmastra* (science of sex-love). But Adi Shankara, being a sannyasi, had no knowledge of this subject; thus, after requesting for some time before entering into this fresh debate, he entered the body of a king by his yogic powers and acquired the knowledge of kmastra. Later, however, Ubhaya Bhrati declined to debate with him and allowed Mandana Mishra to accept sannyasa with the monastic name, Surevarcrya as per the agreed rules of the debate.

***Dig-vijaya:*** Adi Shanara then travelled with his disciples to Maharashtra and Srisailam. In Srisailam, he composed Shivanandalahari, a devotional hymn to Shiva. The Madhaviya Shankaravijayam says that when Shankara was about to be sacrificed by a Kapalika, the god Narasimha appeared to save Shankara on Padmapada's prayer to him.

So Adi Shankara composed the Laksmi-Narasimha stotra to praise Narasimha. He next travelled to Gokarna, the temple of Hari-Shankara and the Mkkambika temple at Kollur. At Kollur he accepted a boy believed to be dumb by his parents, as his disciple. He gave him the name, Hastmalakcrya (one with the amalaka fruit on his palm, *i.e.*, one who has clearly realized the Self). Next, he visited Zangeri to establish the Zrada Pamham and made Tomakcrya his disciple.

After this, Adi Shankara began a *Dig-vijaya* (missionary tour) for the propagation of the Advaita philosophy by controverting all philosophies opposed to it. With King Sudhanva of Kerala as companion, Shankara passed through Tamil Nadu, Andhra Pradesh and Vidarbha. He then started towards Karnataka where he encountered a band of armed Kapalikas. King Sudhanva, with his army, resisted and defeated the kapalikas. They safely reached Gokarna where Shankara defeated in debate the Shaiva scholar, Neelakanta.

Proceeding to the west in Dwarka, Shankara defeated the Vaicnavas in debate. Bhamma Bhskara of Ujjayini, the proponent of Bhedbeda philosophy, was humbled. All the scholars of Ujjayini (also known as Avanti) accepted Adi Shankara's philosophy. He then defeated the Jainas at a place called Bahlika. Later, he had an encounter with a tantrik, Navagupta at Kamarupa. Navagupta pretended to have become a disciple, but later caused to develop a rectal fistula. However, Adi Shankara was soon cured and Navagupta later died of the same disease.

Adi Shankara thus travelled throughout India, from the South to Kashmir and Nepal, preaching to the local populace and debating philosophy with Hindu, Buddhist and other scholars and monks along the way.

***Accession to Sarvajnapitha:*** Adi Shankara visited *Sarvajnapimha* in Kashmir (now in Pakistan-occupied Kashmir). The Madhaviya Shankara Vijaya states this temple had four doors for scholars from the four cardinal directions. The southern door (representing South India) had never been opened, indicating that no scholar from South India had entered the Sarvajna Pitha. Adi Shankara opened the southern door by defeating in debate all the scholars there in all the various scholastic disciplines such as Mimamsa, Vedanta and other branches of Hindu philosophy; he ascended the throne of Transcendent wisdom of that temple. The Madhaviya Shankara Vijaya states that Goddess Saraswati (goddess of knowledge and all literary arts) herself proclaimed the unquestioned scholarly triumph of Adi Shankara on this occasion.

Shankara then travelled to Kedarnath and attained videha mukti (final liberation) at the age of thirty two. However, there are variant traditions on the location of his last days. The Kanchi matha followers believe he achieved *videha mukti* in Kanchi. Another tradition expounded by *Keraliya Shankaravijaya* places his place of death as Vadakkunnathan (Vaccala) temple in Thrissur, Kerala.

***Dates:*** Modern scholarship is agreed on the 788 – 820 CE date, though it has proved impossible to reach agreement on Adi Shankara's precise dates of birth or death. Some Shankara Mamhas, however, ascribe much earlier dates to him. If these dates were true, they would require moving back the date of Buddha (wich serves as an anchor for modern academic history of India). Of the major Shankara Mamhas active today, the ones at Kanchi, Dwaraka, and Puri ascribe the dates 509 – 477 BCE to Adi Shankara. The Zaegeri Zrad Pamham, on the other hand, accepts the 788–820 CE dates. According to Swami Niranjanananda Saraswati's biography of Adi Shankara, published in his book *Sannyasa Darshan*, Adi Shankara was born in Kalady, Kerala, in 686, and attained mahasamadhi at Kedarnath, Uttaranchal, in 718.

***Mathas:*** Adi Shankara founded four Mamhas (abbeys), to guide the Hindu religion. These are at Sringeri in Karnataka in the south, Dwaraka in Gujarat in the west, Puri in Orissa in the east, and Jyotirmath (Joshimath) in Uttaranchal in the north. Hindu tradition states that he put in charge of these mathas his four main disciples: Sureshwaracharya, Hastamalakacharya, Padmapadacharya, and Totakacharya respectively.

The heads of the mathas trace their authority back to these figures. Each of the heads of these four mathas takes the title of Shankaracharya (the learned Shankara) after the first Shankara. The matha at Kanchipuram, Tamil Nadu, claims that it was founded by Adi Shankara.

***Advaita Vedanta in a summary:*** Adi Shankara's treatises on the Upanishads, the Bhagavad Gita and the Brahma Sutras are his principal and almost undeniably his own works. Although he mostly adhered to traditional means of commenting on the Brahma Sutra, there are a number of original ideas and arguments. He taught that it was only through direct knowledge of Brahman that one could be enlightened.

Adi Shankara's opponents accused him of teaching Buddhism in the garb of Hinduism, because his non-dualistic ideals were a bit radical to contemporary Hindu philosophy. However, it may be noted that while the Later Buddhists arrived at a changeless, deathless, absolute truth after their insightful understanding of the unreality of samsara, historically Vedantins never liked this idea. Although Advaita proposes the theory of Maya, explaining the universe as a "trick of a magician", Adi Shankara and his followers see this as a consequence of their basic premise that Brahman alone is real. Their idea of Maya emerges from their belief in the reality of Brahman, rather than the other way around.

***Historical and Cultural Impact:*** At the time of Adi Shankara's life, Hinduism had began to decline because of the influence of Buddhism and Jainism. Hinduism had become divided into innumerable sects, each quarelling with the others. The followers of Mimamsa and Sankhya philosophy were atheists, in so much that they did not believe in God as a unified being. Besides these atheists, there were numerous theistic sects. There were those who rejected the Vedas, like the Charvakas.

Adi Shankara held debates with the leading scholars of all these sects and schools of philosophy to controvert their doctrines. He unified the theistic sects into a common framework of Shanmata system. In his works, Adi Shankara stressed the importance of the Vedas, and his efforts helped Hinduism regain strength and popularity. He travelled on foot to various parts of India to restore the study of the Vedas.

Even though he lived for only thirty-two years, his impact on India and on Hinduism cannot be stressed enough. He reintroduced a purer form of Vedic thought. His teachings and tradition form the basis of Smartism and have influenced Sant Mat lineages. He is the main figure in the tradition of Advaita Vedanta. He was the founder of the Dasanmi Sampradya of Hindu monasticism and *bagmata* of Smarta tradition. He introduced the *Pancyatana* form of worship.

Adi Shankara, along with Madhva and Ramanuja, was instrumental in the revival of Hinduism. These three teachers formed the doctrines that are followed by their respective sects even today. They have been the most important figures in the recent history of Hindu philosophy. In their writings and debates, they provided polemics against the non-Vedantic schools of Sankhya, Vaisheshika etc.

Thus they paved the way for Vedanta to be the dominant and most widely followed tradition among the schools of Hindu philosophy. The Vedanta school stresses most on the Upanishads (which are themselves called Vedanta, *End or culmination of the* Vedas), unlike the other schools that gave importance to texts authored by their founders. The Vedanta schools have the belief that the Vedas, which include the Upanishads, are unauthored, forming a continuous tradition of wisdom transmitted orally. Thus the concept of *apaurusheyatva* (being unauthored) came to be the guiding force behind the Vedanta schools. However, along with stressing the importance of Vedic tradition, Adi Shankara gave equal importance to the personal experience of the student. Logic, grammar, Mimamsa and allied subjects form main areas of study in all the Vedanta schools.

A well known verse, recited in the Smarta tradition, in praise of Adi Shankara is:

*Zruti* salute the compassionate abode of the Vedas, Smritis and Puranas known as Shankara Bhagavatpada, who makes the world auspicious.

***Works of Adi Shankara:*** Adi Shankara's works deal with logically establishing the doctrine of Advaita Vedanta as he saw it in the Upanishads. He formulates the doctrine of Advaita Vedanta by validating his arguments on the basis of quotations from the Vedas and other Hindu scriptures. He gives a high priority to *svnubhava* (personal experience) of the student. A large portion of his works is polemical in nature. He directs his polemics mostly against the Sankhya, Bauddha, Jaina, Vaisheshika and other non-vedantic Hindu philosophies.

Traditionally, his works are classified under Bhacya (*commentary*), Prakarana gratha (*philosophical treatise*) and Stotra (*devotional hymn*). The commentaries serve to provide a consistent interpretation of the scriptural texts from the perspective of Advaita Vedanta. The philosophical treatises provide various methodologies to the student to understand the doctrine. The devotional hymns are rich in poetry and piety, serving to highlight the helplessness of the devotee and the glory of the diety. Of his works, the authenticity of the Viveka Chudamani and a few *Bhashyas* are questioned.

Adi Shankara wrote *Bhashyas* on the ten major Upanishads, the Brahma Sutras and the Bhagavad Gita. In his works, he quotes from Shveshvatara, Kaushitakai, Mahanarayana and Jabala Upanishads, among others. *Bhashyas* on Kaushitaki, Nrisimhatapani and Shveshvatara Upanishads are extant but the authenticity is doubtful. Adi Shankara's commentary on the Brahma Sutras is the earliest one available on this topic. However Adi Shankara mentions older commentaries like those of Dravida, Bhartiprapancha and others.

In his Brahma Sutra Bhashya, Adi Shankara cites the examples of Dharmavyadha, Vidura and others who were born with the knowledge of Brahman acquired in previous births. He mentions that the effects cannot be prevented from working on account of their present birth. He states that the knowledge that arises out of the study of the Vedas could be had through the Puranas and the Itihasas. In the Taittiriya Upanishad Bhashya 2.2, he says:

It has been established that everyone has the right to the knowledge (of Brahman) and that the supreme goal is attained by that knowledge alone.

***Ramanuja:*** Sri Ramanuja Acharya (traditionally dated 1017–1137 BC) was an Indian philosopher and is recognized as the most important saint of Sri Vaishnavism. He held the Vishishtadvaita or qualified Non-dualist belief that the world and Brahman were united, like a soul and a body are. His version of Indian Nondualism differed from Adi Shankara's because he acknowledged the existence of differences, and believed that the identity of an object as a part was as important as the unity of the whole. The Vaishnava Theology espoused by Ramanuja posits tht Brahman is not devoid of attributes but is expressed as a personal God, full of infinite good qualities, as Narayana. The Adishesha on whom Lord Ranganatha of Srirangam rests is believed to be Ramanuja.

Ramanuja was born Ilaya Perumal to a smartha brahmin family in the village of Perumbudur, Tamil Nadu, India in 1017 BC. His father was Keshava Somayaji Deekshitar and mother was Kanthimathi in sect of Vadama.From a young age, his intelligence and ability to comprehend highly abstract philosophical points were legendary. He took initiation from Yadavaprakasa, a renowned Advaitic scholar. Though his new guru was highly impressed with his analytical ability, he was quite concerned by how much emphasis Ramanuja placed on bhakti. After frequent clashes over interpretation, Yadavaprakasa decided the young Ramanuja was becoming too much of a threat and plotted a way to kill him. However, Ramanuja's cousin Govinda Bhatta (a favourite of Yadavaprakasa) discovered the plot and helped him escape. An alternative version is that one of Yadavaprakasa's students plotted to kill Ramanuja as a means of pleasing their teacher, but Sri Ramanuja escaped in the afore-mentioned manner. Yadavaprakasa was horrified when learnt about the conspiracy.

After renouncing the life of a house-holder, Ramanuja travelled to Srirangam to meet an aging Yamunacharya, the pre-eminent Vishishtadvaita philosopher of the time. Yamunacharya had died prior to Ramanuja's arrival, but had left three tasks for Ramanuja to carry out.

- Teach the doctrine of Saranagati (surrender) to God as the means to moksha.
- A Visishtadvaita Bhashya should be written for the Brahma Sutras of Vyasa which had previously been taught orally to the disciples of the Visishtadvaita philosophy.
- That the names of Paraara, the author of Vishnu Purafa, and saint Zachakopa should be perpetuated.

Ramanuja pledged to God to do as he had been requested and accepted Yamunacharya as his *Manasika Acharya*. All three tasks were successfully completed.

***Five Acharyas:*** Swami Ramanuja incorporated teachings from 5 different people who he considered to be his acharyas

1. Peria Nambigal who performed his samasrayana
2. Thirukkotiyur Nambigal : who revealed the meaning of Charama slokam to swami on his 18th trip
3. Thirumalai Nambigal : Ramayana
4. Tirumalai Aandaan : Bhagavad Vishayam
5. Thirukachchi Nambigal : The 6 sentences or Perarulalan

***Visishtadvaita Philosophy:*** Ramanuja's philosophy is referred to as Vishishtadvaita because it combines Advaita (oneness of God) with Vishesha (attributes). The philosophy is monotheistic.

***Differences with Sankara:*** Adi Sankara had argued that all qualities or manifestations that can be perceived are unreal and temporary. They are a result of ignorance. Ramanuja believed them to be real and permanent and under the control of the Brahman.

God can be one despite the existence of attributes, because they cannot exist alone; they are not independent entities. They are Prakaras or the modes, Sesha or the accessories, and Niyama or the controlled aspects, of the one Brahman.

In Sri Ramanuja's system of philosophy, the Lord (Narayana) has two inseparable Prakaras or modes, *viz.*, the world and the souls. These are related to Him as the body is related to the soul. They have no existence apart from Him. They inhere in Him as attributes in a substance. Matter and souls constitute the body of the Lord. The Lord is their indweller. He is the controlling Reality. Matter and souls are the subordinate elements. They are termed Viseshanas, attributes. God is the Viseshya or that which is qualified.

History shows that the followers of Sankara are answerless till date to the strong arguments of Ramanuja (in his sri bhashya) and his followers (satadushani of desika). In a bid to escape strong objections raised by Ramanuja and his successors, most advaitins take a disguised route of neo vedantism, where they argue that vaishnavism is one another path to realise brahman. Ironically, the very brahman of Ramanuja and Sankara are different.

Ramanuja opines, wrong is the position of the Advaitins that understanding the Upanishads without knowing and practicing dharma can result in Brahman knowledge. The knowledge of Brahman that ends spiritual ignorance is meditational, not (as Advaitins seem to presume) testimonial or verbal.

In contrast to Ṡankara, Ramanuja holds, There is no knowledge source in support of the claim that there is a distinctionless (homogeneous) Brahman. All knowledge sources reveal objects as distinct from other objects. All experience reveals an object known in some way or other beyond mere existence. Testimony depends on the operation of distinct sentence parts (words with distinct meanings). Thus the claim that testimony makes known that reality is distinctionless is contradicted by the very nature of testimony as a knowledge means. Even the simplest perceptual cognition reveals something (Bessie) as qualified by something else (a broken hoof, "Bessie has a broken hoof," as known perceptually). Inference depends on perception and makes the same distinct things known as does perception.

Against the Advaita contention that perception cannot make known distinctness but only homogeneous being since distinctness cannot be defined, well, sorry, perception makes known generic characters (cowhood and the like) that differentiate things. If what you Advaitins say were true, why should not a person looking for a horse be satisfied with a buffalo? Remembering could not be distinguished from perceiving, because there would be only the one object (being). And no one would be deaf or blind. Furthermore, Brahman would be an object of perception and the other sources (prameya).

He also holds, The Advaitin argument about prior absences and no prior absence of consciousness is wrong. Similarly the Advaitin understanding of a-vidya (not-Knowledge), which is the absence of spiritual knowledge, is incorrect. "If the distinction between spiritual knowledge and spiritual ignorance is unreal, then spiritual ignorance and the self are one."

***The Seven Objections to Shankara's Advaita:*** Ramanuja picks out what he sees as seven fundamental flaws in the Advaita philosophy for special attack: he sees them as so fundamental to the Advaita position that if he is right in identifying them as involving doctrinal contradictions, then Sankara's entire system collapses. He argues:

I. The nature of Avidya. Avidya must be either real or unreal; there is no other possibility. But neither of these is possible. If Avidya is real, non-dualism collapses into dualism. If it is unreal, we are driven to self-contradiction or infinite regress.

II. The incomprehensibility of Avidya. Advaitins claim that Avidya is neither real nor unreal but incomprehensible, (anirvacaniya). All cognition is either of the real or the unreal: the Advaitin claim flies in the face of experience, and accepting it would call into question all cognition and render it unsafe.

III. The grounds of knowledge of Avidya. No pramana can establish Avidya in the sense the Advaitin requires. Advaita philosophy presents Avidya not as a mere lack of knowledge, as something purely negative, but as an obscuring layer which covers Brahman and is removed by true Brahma-vidya. Avidya is positive nescience not mere ignorance. Ramanuja argues that positive nescience is established neither by perception, nor by inference, nor by scriptural testimony. On the contrary, Ramanuja argues, all cognition is of the real.

IV. The locus of Avidya. Where is the Avidya that gives rise to the (false) impression of the reality of the perceived world? There are two possibilities; it could be Brahman's Avidya or the individual soul's (jiva). Neither is possible. Brahman is knowledge; Avidya cannot co-exist as an attribute with a nature utterly incompatible with it. Nor can the individual soul be the locus of Avidya: the existence of the individual soul is due to Avidya; this would lead to a vicious circle.

V. Avidya's obscuration of the nature of Brahman. Sankara would have us believe that the true nature of Brahman is somehow covered-over or obscured by Avidya. Ramanuja regards this as an absurdity: given that Advaita claims that Brahman is pure self-luminous consciousness, obscuration must mean either preventing the origination of this (impossible since Brahman is eternal) or the destruction of it-equally absurd.

VI. The removal of Avidya by Brahma-vidya. Advaita claims that Avidya has no beginning, but it is terminated and removed by Brahma-vidya, the intuition of the reality of Brahman as pure, undifferentiated consciousness. But Ramanuja denies the existence of undifferentiated (nirguna) Brahman, arguing that whatever exists has attributes: Brahman has infinite auspicious attributes. Liberation is a matter of Divine Grace: no amount of learning or wisdom will deliver us.

VII. The removal of Avidya. For the Advaitin, the bondage in which we dwell before the attainment of Moksa is caused by Maya and Avidya; knowledge of reality (Brahma-vidya) releases us. Ramanuja, however, asserts that bondage is real. No kind of knowledge can remove what is real. On the contrary, knowledge discloses the

> real; it does not destroy it. And what exactly is the saving knowledge that delivers us from bondage to Maya? If it is real then non-duality collapses into duality; if it is unreal, then we face an utter absurdity.

He was critical of the caste system. He said, "Does the wearing of a sacred thread make one a Brahmin? One who is devoted to God (Narayana) alone is a Brahmin." His Sarangati philosophy emphasises that anyone, irrespective of colour, creed, caste, sex and religion can surrender their mind, body and soul to the Lotus foot of Lord Narayana and the God would accept him/her. Cited from Sri Ramanuja, His Life, Religion, and Philosophy, published by Sri Ramakrishna Math, Chennai, India.

***Writings:*** Ramanuja's most famous work is known as the Sri Bhasya. It is a commentary on the Brahma Sutras.

Gadhya Thrayam (three compositions)-Vaikunta, Sriranga and Saranagati Gadhyam are great works in Vaishnava philosophy.

His other works are:

- Vedanta Sara (essence of Vedanta)
- Vedanta Sangraha (a resume of Vedanta)
- Vedanta Deepa (the light of Vedanta).

An interesting point in Ramanuja's works is that, He happens to have composed all his works only in the Sanskrit language.

***Brindavan:*** Ramanuja's thiruvarasu (tomb) is located in Srirangam, Tamil Nadu within the temple complex, where he attained his Acharyan Thiruvadi (the lotus foot of his Acharya). His Thirumeni is covered with chandan (sandalwood paste) and saffron, is open to the public.

***Madhvacharya:*** Madhvacharya (1238-1317) was the chief proponent of Tattvavda (True Philosophy), popularly known as Dvaita or dualistic school of Hindu philosophy. It is one of the three most influential Vedanta philosophies. Madhva was one of the important philosophers during the Bhakti movement. He was a pioneer in many ways, going against standard conventions and norms. Madhvacharya is believed by his followers to be the third incarnation of Vayu, aka Mukhyaprana, after Hanuman and Bhima.

***Formative Years:*** Shri Madhvacharya was born as *Vaasudeva* to Madhyageha Bhatta (father) and Vedavati (mother) at Pajaka in Udupi, Karnataka. The basic tenet of Madhva philosophy is the existence of two kinds of realities, namely "Independent reality" (or svatantra tatva) and "Dependent reality" (or asvatantra tatva).

*Independent reality* (svatantra)-who/which e refer to as "God". 5 generic names are applicable to God (as per shreemad bhaagavata mahaapuraan). They are "brahman", "para brahman", "aatman", "paramaatman" and "bhagavaan". It is to be noted that by "brahman", aacharya madhva does not mean "chaturmukha brahma". The word "brahman" refers to "brihat" (meaning "big") *i.e.*, the one who/which is the biggest in this universe which, is none other than "God".

In general, aachaarya madhva's important message is that every word, every sound in this entire universe only means God which he equated with Vishnu. According to Madhva, only Brahman is indepedent in every sense of the word.

*Dependent reality* (asvatantra) plurality of Jivas and Prakriti or Nature. Both the Jivas and Nature are dependent on Brahman for their very "being" and "becoming". This dependence is expressed metaphorically as *Bimba-pratibimba* (source-reflection) relation. The reflection is in every way dependent on the source that gets reflected.

***Madhva's Credo:*** Madhva was motivated by his four convictions-

1. A determination to remain true to experience above everything.
2. A commitment to sound reasoning
3. A fervent devotion to a personal God (Vishnu) that drove all his actions
4. Fearless tenacity in expounding his vision in the most hostile environments.

***Philosophy of Realism:*** Dvaita school belongs to the Realist school of Indian philosophy, in the same category as Samkhya, Nyaya, Vaisheshika and Purva mimamsa schools. They believe that the universe is a real creation of Brahman. The plurality of souls are bound by a "real" bondage due to beginning-less ignorance, and sadhana through Vishnu bhakti is the only way to be released from this bondage. Further, Madhva explains that Jnana or knowledge alone is not sufficient for the release from beginningless avidya or ignorance, since this bondage is sustained by the "Will" of Brahman and so needs Vishnu Prasadam to ultimately break the bonds of My.

***Five Differences:*** Dvaita, or Dualistic philosophy (known severally as Bheda-vada, Tattva-vada, and Bimba-pratibimba-vada), asserts that the difference between the individual soul or jiva, and God, (Ishvara or Vishnu), is eternal and real. Actually, this is just one of the five differences that are so stated — all five differences that constitute the universe are eternal.

The five are given by:

*jiiveshvara bheda chaiva jadeshvara bheda tatha*

*jiiva-bhedo mithashchaiva jada-jiiva-bheda tatha*

mithashcha jada-bhedo.ayam prapajncho bheda-panchakah-paramashruti

"The difference between the jiva (soul) and Ishvara (Creator), and the difference between jada (insentient) and Ishvara; and the difference between various jivas, and the difference between jada and jiva; and the difference between various jadas, these five differences make up the universe." From the Paramopanishad a.k.a. Parama-shruti, as quoted by Ananda Tirtha in his 'Vishnu-tattva-vinirnaya'. Another way of saying this is that these five fundamental real differences are between: Selves and Brahman; matter and Brahman; one Self and another Self; matter and Selves; and, matter and matter.

Contrary to the Idealistic schools like Yogacara, Madhyamika buddhism or Advaita, Dvaita maintains that difference is in the very nature of a substance. This is the reason why some refer to the doctrine of Tattvavada (the preferred name) as 'Dvaita'. However, 'Dvaita' is thought to be inadequately representative of the true grain of Tattvavada.

The doctrine of Tattvavada is considered to be eternal (in a flow-like sense, just as Creation is eternal); in historical times, it was revived by Ananda Tirtha, who is also known as Madhvacharya. Because of this, followers of Tattvavada are called Madhvas, meaning followers of Madhva.

Souls are not created by God:

*Madhvacharya has established that souls are eternal, and are not created by God as in the Semitic religions. The souls are dependent, not generated by Vishnu but coexist with Him eternally, supported by His will and entirely controlled by Him. The souls, however, are dependent on Him in their pristine nature and in all transformation that they may undergo.*

Three Classes of Souls and Eternal Damnation:

*Additionally, Madhvacharya differed significantly from traditional Hindu beliefs in his concept of eternal damnation. For example, he divides souls into three classes, one class which qualify for liberation, Mukti-yogyas, another subject to eternal rebirth or eternally transmigrating due to samsara, Nitya-samsarins, and significantly, a class that is eventually condemned to eternal hell or Andhatamas, known as Tamo-yogyas.*

Madhva followers cite authorities such as Bhagavad Gita, Chapter 17, verses 2 et seq. "'There are three types of inclination, which are the self-same natures of the souls, these being satvika, rajasa, and tamasa," Chapter 16, verses 19-20, "These cruel haters, worst among men in the world, I hurl these evil-doers into the wombs of demons only. Entering into demoniacal wombs and deluded, birth after birth, not attaining me, they thus fall, Oh Arjuna, into a condition still lower than that," for their concept of eternal damnation. Madhvacharaya was the first in the recent years who revived the timeless Vaishnava tradition. There were 21 different Bashayas (commenteries) before Sri Madhvacharaya. He is the first to establish the facts of tri-patriate classification of souls. By contrast, most Hindus believe that souls will eventually obtain moksha, even after millions of rebirths. Madhvacharya's Theology as an Answer to the Eternal Problem of Evil:

> *By following the concepts of souls not being created by God and classification of the souls, Madhvacharya provides a lucid answer to the problem of evil by seeking a root cause like the intrinsic nature of the soul itself. Often, evil behaviour displayed in the world might not be just the nature of the soul but also depends upon the timeless actions (Karma) of the soul itself.*

***Impact of Dvaita Movement:*** Madhva's Dualistic view, along with Shankara's Advaita or Nondualism and Ramanuja's Qualified Nondualism or Vishishtadvaita form some core Indian beliefs on the nature of reality.

Madhva is considered to be one of the influential theologians in Hindu history. He revitalized an Hindu monotheism in light of attacks, theological and physical, by foreign invasion. Great leaders of the Vaishnava Bhakti movement, in Karnataka, for example, Purandara Dasa and Kanaka Dasa were part of the Dvaita traditions. Also, the famous Hindu saint, Raghavendra Swami, was a leading figure in the Dvaita tradition.

Madhvacharaya during his time not only established dvaita philosophy, but also displayed extraordinary strength and skills to show that he is the third avatara of Vayu, who came down to earth to help people suffering from delusional philosophies and guide them in the right path. Madhvacharaya at the age of 79, year 1317, disappeared from the eyes of humans and continue to reside in Upper Badari in his continuing service to his eternal master Sri Vedavyasa.

Narayana pandit Acharya captures Madhvacharaya's life in a beautiful poetic verses in his "Shri Madhva Vijaya" which is in 16 Sarga (chapters), this book is an authentic work composed during his own time. This is a very rare work, there is no evidence of anyone composing works on any major philosopher like this before or after him from other disciplines. "Shri Madhva Vijaya" is a composition which captures life history of Mahdvaharaya.

***Religious Establishments:*** The main icon (vigraha) in Udupi of Lord Krishna was established by Madhvacharya. The 8 monasteries (ashta mathas) of Udupi have been following his philosophy since then.

***Comparison to Mainstream Hinduism:*** The teachings of Sri Madhvacharya were in many ways quite radical for his times. One example is his doctrine of eternal damnation. This idea which is prevalent in Abrahamic religions, is generally not endorsed by most schools of Hindu philosophy. Many Western scholars see this and the importance given to *Mukhyaprana*-as the mediator to Brahman-as Christian influence. But Dvaita scholars argue that Madhva has derived these concepts from within the Vedic framework. There are many instances in the Upanishads and Gita that support Madhva's position. They also argue that since the knowledge of whether a Jiva is *Muktiyogya* (liberation-worthy) or *Tamoyogya* (damnation-worthy) is not accessible to the Jiva himself, this philosophy does not discourage sadhana for anyone.

***Place of Birth :*** Pajaka, Udupi

***Guru :*** Achyuta Prajnya

***Names :*** **1. Vasudeva**, as named by his parents

**2. Poornaprajnya** (One who knows everything)-named after attaining sainthood

**3. Anandateertha** (One who brings bliss through his preachings)-named after he succeeded his guru

**4. Shri Madhvacharya** This name is not given to him by any one but it comes from the vedas. There is a hymn in the vedas called "balitthaa sookta" where the term "madhva" has been used to refer to "mukhya praana"

***Avatars:***

1. Hanuman
2. Bhima Madhvaachaarya declares, in his work "Vishnu-tatva-vinirnaya:", that he was the one who took the avatars of Hanuman and Bhima. Significantly, the only other person who openly makes such a declaration about his original form is Sri Krishna.

2

# The Vedic Literature

## Rigveda

The *Rigveda* is a collection of Vedic Sanskrit hymns dedicated to the gods. It is counted as the holiest among the series of four Hindu canonical texts known as the Vedas. Geographical and ethnological passages in the Rigveda provide evidence that the Rigveda was composed between 1700–1100 BCE (the early Vedic period) in the Punjab (Sapta Sindhu) region of the Indian subcontinent. The Rigveda is the oldest of all known religious books, and the oldest book in Vedic Sanskrit or any Indo-European language. The composition of the Rigveda is conventionally dated to before 1500 BC. Some writers have traced astronomical references in the *Rigveda* dating it to as early as 4000 BC, a date well within the late Mehrgarh culture.

Today, this text is revered by Hindus around the world, primarily in India and Nepal. Its verses are recited at prayers, religious functions and other auspicious occasions.

***Text:*** The Rigveda consists of 1,028 hymns (or 1,017 discounting the apocryphal *valakhilya* hymns 8.49–8.59) composed in Vedic Sanskrit, many of which are intended for various sacrifical rituals. This long collection of short hymns is mostly devoted to the praise of the gods. It is organized in 10 books, known as Mandalas. Each Mandala consists of hymns, called *sukta*, which in turn consist individual verses called, plural. The Mandalas are by no means of equal length or age: The "family books", mandalas 2-7, are considered the oldest part of the Rigveda, being the shortest books, arranged by length, accounting for 38% of the text. RV 8 and RV 9, likely comprising hymns of mixed age, account for 15% and 9%, respectively. RV 1 and RV 10, finally, are both the latest and the longest books, accounting for 37% of the text.

***Preservation:*** The *Rigveda* is preserved by two major shakhas ("branches", *i.e.* schools or recensions), *Uakala* and *Backala*. Considering its great age, the text is spectacularly well-preserved and uncorrupted, the two recensions being practically identical, so that scholarly editions can mostly do without a critical apparatus. Associated to *Uakala* is the Aitareya-Brahmana. The *Backala* includes the Khilani and has the Kausitaki-Brahmana associated to it.

This compilation or redaction included the arrangement in books as well as orthoepic changes, such as regularization of sandhi (called by Oldenberg *orthoepische Diaskeunase*). It took place centuries after the composition of the earliest hymns, about coeval to the redaction of the other Vedas.

From the time of its redaction, the text has been handed down in two versions: The *Samhitapatha* has all Sanskrit rules of sandhi applied and is the text used for recitation. The Padapatha has each word isolated in its pausa form and is used for memorization. The Padapatha is, as it were, a commentary to the *Samhitapatha*, but the two seem to be about coeval. The original text as reconstructed on metrical grounds (*viz.* "original" in the sense that it aims to recover the hymns as composed by the Rishis) lies somewhere between the two, but closer to the Samhitapatha.

***Organization:*** The most common numbering scheme is by book, hymn and verse (and *pada* (foot) *a*, *b*, *c*..., if required). *e.g.* the first *pada* is

- "Agni I laud, the high priest" and the final *pada* is
- "for your being in good company"

Hermann Grassmann had numbered the hymns 1 through to 1028, putting the *valakhilya* at the end. The entire 1028 hymns of the *Rigveda*, in the 1877 edition of Aufrecht, contain a total of 10,552 verses, or 39,831 padas. The Shatapatha

Brahmana gives the number of syllables to be 432,000, while the metrical text of van Nooten and Holland (1994) has a total of 395,563 syllables (or an average of 9.93 syllables per pada); counting the number of syllables is not straightforward because of issues with sandhi. Most verses are jagati (padas of 12 syllables), trishtubh (padas of 11 syllables), viraj (padas of 10 syllables) or gayatri or anushtubh (padas of 8 syllables).

***Contents:*** The chief gods of the *Rigveda* are Agni, the sacrificial fire, Indra, a heroic god who is praised for having slain his enemy Vrtra, and Soma, the sacred potion, or the plant it is made from. Other prominent gods are Mitra-Varuna and Ushas (the dawn). Also invoked are Savitar, Vishnu, Rudra, Pushan, Brihaspati, Brahmanaspati, as well as deified natural phenomena such as Dyaus Pita (the sky), Prithivi (the earth), Surya (the sun), Vayu (the wind), Apas (the waters), Parjanya (the rain), Vac (the word), many rivers (notably the Sapta Sindhu, and the Saraswati River). Groups of deities are the Ashvins, the Maruts, the Adityas, the Rbhus, the Vishvadevas (the all-gods).

It contains various further minor gods, persons, concepts, phenomena and items, and fragmentary references to possible historical events, notably the struggle between the early Vedic people (known as Vedic Aryans, a subgroup of the Indo-Aryans) and their enemies, the Dasa.

*Rigveda* manuscript in Devanagari, early 19th century:

- Mandala 1 comprises 191 hymns. Hymn 1.1 is addressed to Agni, and his name is the first word of the *Rigveda*. The remaining hymns are mainly addressed to Agni and Indra. Hymns 1.154 to 1.156 are addressed to Vishnu.
- Mandala 2 comprises 43 hymns, mainly to Agni and Indra. It is chiefly attributed to the Rishi *gatsamda aunohotra*.
- Mandala 3 comprises 62 hymns, mainly to Agni and Indra. The verse 3.62.10 has great importance in Hinduism as the Gayatri Mantra. Most hymns in this book are attributed to *viuvamitra gathina*.

- Mandala 4 consists of 58 hymns, mainly to Agni and Indra. Most hymns in this book are attributed to *Vamadeva Gautama*.
- Mandala 5 comprises 87 hymns, mainly to Agni and Indra, the Visvadevas (gods of the world), the Maruts, the twin-deity Mitra-Varuna and the Asvins. Two hymns each are dedicated to Ushas (the dawn) and to Savitar. Most hymns in this book are attributed to the *atri* family.
- Mandala 6 comprises 75 hymns, mainly to Agni and Indra. Most hymns in this book are attributed to the *barhaspatya* family of Angirasas.
- Mandala 7 comprises 104 hymns, to Agni, Indra, the Visvadevas, the Maruts, Mitra-Varuna, the Asvins, Ushas, Indra-Varuna, Varuna, Vayu (the wind), two each to Saraswati (ancient river/goddess of learning) and Vishnu, and to others. Most hymns in this book are attributed to *vasicmha maitravaurgi*.
- Mandala 8 comprises 103 hymns to different gods. Hymns 8.49 to 8.59 are the apocryphal *valakhilya*. Most hymns in this book are attributed to the *kagva* family.
- Mandala 9 comprises 114 hymns, entirely devoted to *Soma Pavamana*, the plant of the sacred potion of the Vedic religion.
- Mandala 10 comprises 191 hymns, to Agni and other gods. It contains the Nadistuti sukta which is in praise of rivers and is important for the reconstruction of the geography of the Vedic civilization and the Purusha sukta which has significance in Hindu tradition. It also contains the Nasadiya sukta (10.129), probably the most celebrated hymns in the west, which deals with creation.

***Rishis:*** Each hymn of the Rigveda is traditionally attributed to a specific rishi, and the "family books" (2-7) are said to have been composed by one family of rishis each. The main families, listed by the number of verses ascribed to them are:

- Angirasas: 3619 (especially Mandala 6)

- Kanvas: 1315 (especially Mandala 8)
- Vasishthas: 1267 (Mandala 7)
- Vaishvamitras: 983 (Mandala 3)
- Atris: 885 (Mandala 5)
- Bhrgus: 473
- Kashyapas: 415 (part of Mandala 9)
- Grtsamadas: 401 (Mandala 2)
- Agastyas: 316
- Bharatas: 170

***Translations:*** The *Rigveda* was translated into English by Ralph T.H. Griffith in 1896. Partial English translations by Maurice Bloomfield and William Dwight Whitney exist. Griffith's translation is good, considering its age, but it is no replacement for Geldner's 1951 translation (in German), the only independent scholarly translation so far. The later translations by Elizarenkova depends heavily on Geldner, but Elizarenkova's translation (in Russian) is valuable in taking into account scholarly literature up to 1990. Partial translations exists in many other languages.

***Hindu Tradition:*** According to Indian tradition, the Rigvedic hymns were collected by Paila under the guidance of Vyasa, who formed the Rigveda Samhita as we know it. According to the Uatapatha Brahmana, the number of syllables in the *Rigveda* is 432,000, equalling the number of muhurtas (1 day = 30 muhurtas) in forty years. This statement stresses the underlying philosophy of the Vedic books that there is a connection (bandhu) between the astronomical, the physiological, and the spiritual. The authors of the Brahmana literature described and interpreted the Rigvedic ritual. Yaska was an early commentator of the *Rigveda*. In the 14th century, Sayana wrote an exhaustive commentary on it. Other *Bhacyas* (commentaries) that have been preserved up to present times are those by Madhava, Skacdasvamin and Veckatamadhava.

***Dating and Historical Reconstruction:*** Geography of the Rigveda, with river names; the extent of the Swat and Cemetery cultures are also indicated.

The *Rigveda* is far more archaic than any other Indo-Aryan text. For this reason, it was in the centre of attention of western scholarship from the times of Max Muller. The *Rigveda* records an early stage of Vedic religion, still closely tied to the pre-Zoroastrian Persian religion. It is thought that Zoroastrianism and Vedic Hinduism evolved from an earlier common religious—Indo-Iranian—culture.

The Rigveda's core is accepted to date to the late Bronze Age, making it the only example of Bronze Age literature with an unbroken tradition. Its composition is usually dated to roughly between 1700–1100 BC. The text in the following centuries underwent pronunciation revisions and standardization (samhitapatha, padapatha). This redaction would have been completed around the 7th century BC.

Writing appears in India around the 5th century BC in the form of the Brahmi script, but texts of the length of the Rigveda were likely not written down until much later, the oldest surviving manuscript dating to the 11th century. While written manuscripts were used for teaching in medieval times, they were written on bark or palm leaves, which decomposed quicker in the tropical climate, until the advent of the printing press from the 16th century. The hymns were thus preserved by oral tradition for up to a millennium from the time of their composition until the redaction of the Rigveda, and the entire Rigveda was preserved in shakhas for another 2,500 years from the time of its redaction until the *editio princeps* by Muller, a collective feat of preservation unparalleled in any other known society.

Puranic literature names Vidagdha as the author of the Pada-text. Other scholars argue that Sthavira Sak of the Aitareya Aranyaka is the padakara of the RV. After their composition, the texts were preserved and codified by a vast body of Vedic priesthood as the central philosophy of the Iron Age Vedic civilization.

The *Rigveda* describes a mobile, nomadic culture, with horse-drawn chariots and metal (bronze) weapons. According to some scholars the geography described is consistent with that of the Punjab (Gandhara): Rivers flow north to south, the mountains are relatively remote but still reachable (Soma is a plant found in the mountains, and it has to be purchased, imported by merchants). Nevertheless, the hymns were certainly composed over a long period, with the oldest elements possibly reaching back before the split of Proto-Indo-Iranian (around 2000 BC). Thus there is some debate over whether the boasts of the destruction of stone forts by the Vedic Aryans and particularly by Indra refer to cities of the Indus Valley civilization or whether they hark back to clashes between the early Indo-Aryans with the BMAC in what is now northern Afghanistan and southern Turkmenistan (separated from the upper Indus by the Hindu Kush mountain range, and some 400 km distant).

In any case, while it is highly likely that the bulk of the *Rigveda* was composed in the Punjab, even if based on earlier poetic traditions, there is no mention of either tigers or rice in the *Rigveda* (as opposed to the later Vedas), suggesting that Vedic culture only penetrated into the plains of India after its completion. Similarly, it is assumed that there is no mention of iron although the term ayas (metal) occurs in the Rigveda. The Iron Age in northern India begins in the 12th century BC with the *Black and Red Ware* (BRW) culture. This is a widely-accepted timeframe for the beginning codification of the *Rigveda* (*i.e.* the arrangement of the individual hymns in books, and the fixing of the samhitapatha (by applying Sandhi) and the padapatha (by dissolving Sandhi) out of the earlier metrical text), and the composition of the younger Vedas. This time probably coincides with the early Kuru kingdom, shifting the centre of Vedic culture east from the Punjab into what is now Uttar Pradesh.

Some of the names of gods and goddesses found in the *Rigveda* are found amongst other belief systems based on Proto-Indo-European religion as well: Dyaus-Pita is cognate with Greek Zeus, Latin Jupiter (from deus-pater), and Germanic Tyr; while Mitra is cognate with Persian *Mithra*; also, Ushas with Greek Eos and Latin Aurora; and, less certainly, Varuna with Greek Uranos. Finally, Agni is cognate with Latin *ignis* and Russian *ogon*, both meaning "fire". Some writers have traced astronomical references in the *Rigveda* dating it to as early as 4000 BC, a date well within the Indian Neolithic. Claims of such evidence remain controversial, but are a key factor in the development of the Proto-Vedic Continuity theory.

N. Kazanas in a polemic against the "Aryan Invasion Theory" suggests a date as early as 3100 BC, based on an identification of the early Rigvedic Saraswati River as the Ghaggar-Hakra and on glottochronological arguments. Being a polemic against mainstream scholarship, this is in diametral opposition to views in mainstream historical linguistics, and supports the controversial Out of India theory, which assumes a date as late as 3000 BC for the age of late Proto-Indo-European itself.

***Flora and Fauna in the Rigveda:*** The horse (Ashva) and cattle play an important role in the Rigveda. There are also references to the elephant (Hastin, Varana), Camel (Ustra, especially in Mandala 8), Buffalo (Mahisa), lion (Simha) and to the Gaur in the Rigveda. The peafowl (Mayura) and the Chakravaka (*Anas casarca*) are birds mentioned in the Rigveda.

### *Indian Views of Recent Past*

The Hindu perception of the *Rigveda* has moved away from the original ritualistic content to a more symbolic or mystical interpretation. For example, instances of animal sacrifice are not seen as literal slaughtering but as transcendental processes. The Rigvedic view is seen to consider the universe to be infinite in size, dividing knowledge into two categories: *lower* (related to objects, beset with paradoxes) and *higher* (related to the perceiving subject, free of paradoxes). Swami Dayananda, who started the Arya Samaj and Sri Aurobindo have emphasized a spiritual (adhyatmic) interpretation of the book.

The Saraswati river, lauded in RV 7.95 as the greatest river flowing from the mountain to the sea is sometimes equated with the Ghaggar-Hakra river, which went dry perhaps before 2600 BC or certainly before 1900 BC. Others argue that the Saraswati was originally the Helmand in Afghanistan. These questions are tied to the debate about the Indo-Aryan migration (termed "Aryan Invasion Theory") vs. the claim that Vedic culture together with Vedic Sanskrit originated in the Indus Valley Civilisation (termed "Out of India theory"), a topic of great significance in Hindu nationalism, addressed for example by Amal Kiran and Shrikant G. Talageri. Subhash Kak has claimed that there is an astronomical code in the organization of the hymns.

Bal Gangadhar Tilak, also based on astronomical alignments in the Rigveda, in his "The Orion" (1893) claimed presence of the Rigvedic culture in India in the 4th millennium BC, and in his "Arctic Home in the Vedas" (1903) even argued that the Aryans originated near the North Pole and came south during the Ice Age

***Rigvedic Deities:*** There are 1028 hymns in the *Rigveda*, most of them dedicated to specific deities.

*Indra*, the heroic god and slayer of Vrtra and Vala, liberator of the cows and the rivers, *Agni*, the sacrificial fire and messenger of the gods, and *Soma*, the ritual drink dedicated to Indra are the most prominent deities by far. Invoked in groups are the *Vishvedevas* (the "all-gods"), the *Maruts*, violent storm gods in Indra's train and the *Ashvins*, the twin horsemen.

There are two major families of gods, the *Devas* and the *Asuras*. Unlike in later Hinduism, the Asuras are not yet demonized, *Mitra* and *Varuna* being their most prominent members. Aditi is the mother both of Agni and of the *Adityas*, a group of Asuras, led by Mitra and Varuna, with Aryaman, Bhaga, Daksha, Ansa and Savitar.

Surya is the personification of the Sun, but Savitar, the Ashvins and the Rohus, semi-divine craftsmen, also have aspects of solar deities. Other natural phenomena deified include Vayu, (the wind), Dyaus and Prithivi (Heaven and Earth), Dyaus continuing Dyeus, the chief god of the Proto-Indo-European religion, and Ushas (the dawn), the most prominent goddess of the Rigveda, and Apas (the waters).

Rivers play an important role, deified as goddesses, most prominently the Sapta Sindhu and the Saraswati River.

Yama is the mythical first ancestor, also worshipped as a deity, and the god of the underworld and death.

Vishnu and Rudra, the prominent deities of later Hinduism (Rudra being an early form of Shiva) are already present as marginal gods.

The names of Indra, Mitra, Varuna and the Nasatyas are also attested in a Mitanni treaty, suggesting that the religion of the Indo-Aryan Mitanni ruling class was very close to that of the Rigveda.

***Deities by Prominence:*** List of Rigvedic deities by number of dedicated hymns, after Griffith (1888). Some dedications are to twin-deities, such as Indra-Agni, Mitra-Varuna, Soma-Rudra, here counted doubly.

- Indra 289
- Agni 218
- Soma 123 (most of them in the Soma Mandala)
- Vishvadevas 70
- The Asvins 56
- Varuna 46
- the Maruts 38
- Mitra 28
- Ushas 21
- Vayu (Wind) 12
- Savitar 11
- The Rohus 11
- Pushan 10
- The Apris 9
- Brhaspati 8
- Surya (Sun) 8
- Dyaus and Prithivi (Heaven and Earth) 6, plus 5.84 dedicated to Earth alone
- Apas (Waters) 6
- Adityas 6
- Vishnu 6
- Brahmanaspati 6
- Rudra 5

- Dadhikras 4
- The Saraswati River/Saraswati 3
- Yama
- Parjanya (Rain) 3
- Vac (Speech) 2 (mentioned 130 times, deified *e.g.* in 10.125)
- Vastospati 2
- Vishvakarman 2
- Manyu 2
- Kapinjala (the Heathcock, a form of Indra) 2

Minor deities (one single or no dedicated hymn)

- Manas (Thought), prominent concept, deified in 10.58
- Dakshina (Reward), prominent concept, deified in 10.107
- Jnanam (Knowledge), prominent concept, deified in 10.71
- Purusha ("Cosmic Man" of the Purusha sukta 10.90)
- Aditi
- Bhaga
- Vasukra
- Atri
- Apam Napat
- Ksetrapati
- Ghrta
- Nirrti
- Asamati
- Urvasi
- Pururavas
- Vena
- Aranyani
- Mayabheda
- Tarksya
- Tvastar
- Saranyu

## Anukramani

An ***Anukramani*** is an index of Vedic hymns, recording poetic meter, content, and traditions of authorship.

The most important Anukramani is Katyayana's *Sarvanukramani* of the Rigveda (ca. 2nd century BC), recording the first word, the number of verses, name and family of poets (Rishis), names of deities and metres for each of the 1,028 hymns of the Rigveda.

Mayrhofer (2003) discusses the personal names contained in the Rigveda Anukramani, counting 543 items. Academic opinion regarding the age and authenticity of the tradition of these names is not unanimous. Mayrhofer suggests that Hermann Oldenberg (1888) was essentially correct in assuming that "the editors of the lists of authors [...] [possessed] a correct notion of the families associated with these Mandalas [the Rigvedic "family books" 2–7], *possibly* rooted in tradition. Beyond this, they do not betray as much as the slightest sign of any genuine tradition on the hymn authors" (p. 229).

## Yajurveda

The ***Yajurveda*** is one of the four Hindu Vedas. It contains religious texts focusing on liturgy, rituals and sacrifices, and how to perform the same.

***Collections:*** There are two primary collections or *samhitas* of the Yajurveda: Shukla (white) and Krishna (black). Both contain the verses necessary for rituals, but the Krishna Yajurveda has additional prose commentary and detailed instructions within the work itself.

***Shukla Yajurveda:*** There are two (nearly identical) *shakhas* or recensions of the Shukla (White) Yajurveda, both known as Vajasaneyi-Samhita (VS):

- *Vajasaneyi Madhyandiniya* (VSM)
- *Vajasaneyi Kanva* of Kosala (VSK)

The former is popular in North India, Gujarat, parts of Maharashtra (north of Nasik) and northern parts of Orissa, and thus commands a numerous following. The Kanva Shakha is popular in Maharashtra (south of Nasik), most of Orissa, Karnataka, Andhra Pradesh and Tamil Nadu. Sureshvaracharya, one of the four main disciples of Jagadguru Adi Shankara, is said to have followed the Kanva *shakha*. The Guru himself followed the Taittiriya Shakha with the Apastamba Kalpasutra.

The Vedic rituals of the Ranganatha-swamy Temple at Srirangam, the second biggest temple in India, are performed according to the Kanva *shakha*. The White Yajurveda has two Upanishads associated with it: the Isa Vasya and the Brihadaranyaka upanishads. The Brihadaranyaka Upanishad is the most voluminous of all Upanishads.

The VS has forty chapters or *adhyayas*, containing the formulas used with the following rituals:

1-2: New and Full Moon sacrifices
3: Agnihotra
4-8: Somayajna
9-10: Vajapeya and Rajasuya, two modifications of the Soma sacrifice
11-18: Construction of altars and hearths, especially the Agnicayana
19-21: Sautramani, a ritual originally counteracting the effects of excessive Soma-drinking
22-25: Ashvamedha
26-29: Supplementary formulas for various rituals
30-31: Purushamedha
32-34: Sarvamedha
35: Pitriyajna
36-39: Pravargya
40: The final adhyaya is the famous Isha Upanishad

### *Krishna Yajurveda*

There are four recensions of the Krishna ("black" or "dark") Yajurveda:

- *taittiriya sachita* (TS) of Panchala
- *maitrayani sachita* (MS)

- *caraka-katha sachita* (KS) of Madra
- *kapicmhala-katha sachita* (KapS) of Bahika

Each of the recensions has a *Brahmana* associated with it, and some of them also have associated Shrautasutras, Grhyasutras, Aranyakas, Upanishads and Pratishakhyas.

***The Taittiriya Shakha:*** The best known of these recensions is the TS, named after Tittiri, a pupil of Yaska. It consists of 8 books or *kandas*, subdivided in chapters or *prapathakas*, further subdivided into individual hymns. Some individual hymns in this Samhita have gained particular importance in Hinduism; *e.g.* TS 4.5 and TS 4.7 constitute the Shri Rudram Chamakam, while 1.8.6.i is the Shaivaite Tryambakam mantra. The formula *bhur bhuva, suva,* prefixed to the (rigvedic) Gayatri mantra is also from the Yajurveda, appearing four times. The Taittiriya recession of the Black Yajurveda *shakha* most prevalent in southern India. Among the followers of this Shakha, the Apastamba Sutras is the common Shrautasutra associated with the Shakha.

The Taittiriya Shakha consists of Taittiriya Samhita (having seven kaandas), Taittiriya Braahmana (having three kaandas), Taittiriya Aaranyaka (having seven prashnas), Taittiriya Upanishad (having three *prashnas* or *vallis*-Sheeksha valli, Ananda valli and Bhrigu valli) and the Mahanarayana Upanishad. The Taittiriya Upanishad and Mahaanaarayana Upanishad are considered to be the seventh, eighth, ninth and tenth prashnas of the Aaranyaka. The words *prapaathaka* and *kaanda* (meaning sections) are interchangeably used in the Vedic literature. *Prashna* and *valli* refer to sections of the Aaranyaka.

***Legend:*** Legend has it that the vedic seer Yajnavalkya studied the Yajurveda collection under the tutelage of sage Vaishampayana maternal uncle of Yajnavalkya. Yajnavalkya's birth was with a purpose as purported by Gods. He was an 'Ekasandhigraahi', meaning he learnt anything with just once teaching. The two came to have serious differences in interpretation. On one occasion, Vaishampayana was so enraged that he demanded the return of all the knowledge he has imparted to Yajnavalkya. Yagnavalkya regurgitated all the knowledge he had learnt in the form of flesh.

The other disciples of Vaishampayana, eager to receive this knowledge, assumed the form of *tittiriya* birds and ate that flesh. Thus, that knowledge came to be called the Taittiriya Samhita (from the word *tittiriya*). There is a secret meaning in this. As we know, it is impossible to regurgiate knowledge or learning, but the words are used in 'srimad bhaagavata'.

It is also of interest to know why the Guru himself did not take the vedaas he taught Yajnyavalkya, but instead asked his deciples to eat the flesh. Any way, after having regurgitated out the knowledge acquired from his teacher, Yajnavalkya worshipped Surya (the Sun God) and acquired new knowledge directly from God Narayana (suryantargata narayana) who preached Shukla yajurveda by taking vaaji rupa, (god with horse face) which was compiled to become the *Shukla Yajurveda*. This is recorded in srimad bhaagavata.

***Large Numbers:*** The Yajurveda documents the earliest known use of numbers up to a trillion (*parardha*). It even discusses the concept of numeric infinity (*purna* "fullness"), stating that if you subtract *purna* from *purna*, you are still left with *purna*.

## Samaveda

The *Samaveda* is third in the usual order of enumeration of the four Vedas, the ancient core Hindu scriptures.

The Samaveda ranks next in sanctity and liturgical importance to the Rigveda or Veda of Recited praise. Its Sanhita, or metrical portion, consists chiefly of hymns to be chanted by the Udgatar priests at the performance of those important sacrifices in which the juice of the Soma plant, clarified and mixed with milk and other ingredients, was offered in libation to various deities. The Collection is made up of hymns, portions of hymns, and detached verses, taken mainly from the Rigveda, transposed and rearranged, without reference to their original order, to suit the religious ceremonies in which they were to be employed.

The verses are not intended to be chanted, but to be sung in specifically indicated melodies using the seven *svaras* or notes. Such songs are called *Samagana* and in this sense the Samaveda is really a book of hymns.

In these compiled hymns there are frequent variations, of more or less importance, from the text of the Rigveda as we now possess it which variations, although in some cases they are apparently explanatory, seem in others to be older and more original than the readings of the Rigveda. In singing, the verses are still further altered by prolongation, repetition and insertion of syllables, and various modulations, rests, and other modifications prescribed, for the guidance of the officiating priests, in the Ganas or Song-books.

Two of these manuals, the Gramageyagana, or Congregational, and the Aranyagana or Forest Song-Book, follow the order of the verses of part I, of the Sanhita, and two others, the Uhagana, the Uhyagana, of Part II. This part is less disjointed than part I, and is generally arranged in triplets whose first verse is often the repetition of a verse that has occurred in part I.

The Samaveda survives in a single shakha or recession, the Kauthuma shakha, with a second shakha, Jaiminiya (or Talavakara), surviving fragmentarily, the Jaiminiya Samhita. From the Jaiminiya shakha, we also have the Jaiminiya Brahmana, the Jaiminiya Upanishad Brahmana and the Kena Upanishad.

Since the Samaveda is written in verse it can be sung. This decade has seen a poetic translation of Samveda in Hindi. This translation was done by Dr. Mridul Kirti and is called "Samveda Ka Hindi Padyanuvad"

Samaveda had originally 1000 shakhas. Here are the names of some of them:

- Ranayana
- Shatyamukhya
- Vyasa
- Bhaguri

- Oulundi
- Goulgulvi
- Bhanuman-oupamayava
- Karati
- Mashaka Gargya
- Varsgagavya
- Kuthuma
- Sgakugitra
- Jaimini

One of the leading exponents of Samaveda was Pandit Rewashankar Shastri of Padadhari, Gujarat India. He was awarded with the title of 'Samaveda Martanda' by His Holiness Paramacharya Sankaracharya of Kanchi Kamakoti Peetam. This honour was conferred on Pt. Rewashankar Shastri in Madras, India in late 1960s.

## Atharvaveda

The *Atharvaveda* (Sanskrit, *Atharvaveda*, a tatpurusha compound of *atharvan*, a type of priest, and *veda* meaning "knowledge") is a sacred text of Hinduism, and one of the four Vedas, often called the "fourth Veda". According to tradition, the Atharvaveda was mainly composed by two groups of rishis known as the Bhrigus and the Angirasas. Additionally, tradition ascribes parts to other rishis, such as Kauuika, Vauicmha and Kashyapa. There are two surviving recensions (uakhas), known as *aunakiya* (*AVS*) and *Paippalada* (*AVP*).

***Status:*** The Atharvaveda, while undoubtedly belonging to the core Vedic corpus, in some ways represents an independent parallel tradition to that of the Rigveda and Yajurveda.

The Jaina and Buddha texts are considerably more hostile to the AV (they call it Ạggvana or Ahavana Veda) than they are to the other Hindu texts. The Atharva Veda is less predominant than other Vedas, also the Gayatri mantra used in Atharva Veda is different from other 3 Vedas a special initiation of Gayatri is required to learn the Atharva Veda. The Hindus believe the mantras are highly powerful, the Atharvan

Pariihthas (appendices) themselves state that specific priests of the Mauda and Jalada schools should be avoided or strict discipline should be followed as per the rules and regulations laid by the Atharva Veda. It is even stated that women associated with Atharvan may suffer from abortions if pregnant women remain while the chants for warfare are pronounced.

***Recensions:*** The *Caraavyuha* (attributed to Shaunaka) lists nine shakhas or Schools of the Atharvaveda:

1. Paippalada
2. stauda
3. mauda
4. aunakiya
5. jajala
6. jalada
7. brahmavada
8. devadarea
9. charaavidya

Of these, only the Ceaunakiya (AVS) and the Paippalada (AVP) recensions have survived. The core Paippalada text is considered earlier than the Ceaunakiya, but both also contain later additions and corruptions. In places where the Ceaunakiya and the Paippalada agree, it is likely the original version. Often, the two recensions in corresponding hymns have a different verse order, or either has additional verses missing from the other.

Additionally, from the Vicgu and Vayu Puranas (older Hindu texts on the gods, goddesses and their histories) it may be possible to glean a few more ancient schools that were not listed in the *Caragavyuha*.

These are:

- sumantu
- kabandha
- kumuda
- uaulkayana
- babhravya

- munjakeœa
- saindhavayana
- nakœatrakalpa
- uantikalpa
- sahitavidhi

At least some of these may have evolved into the other schools mentioned in the Caraavyuha list. Sahitavidhi, Uantikalpa and Nakoeatrakalpa are the five *kalpa* texts adduced to the Ceaunakiya tradition and not separate schools of their own.

There are two main circum-vedic texts associated with the AV, the vaitana sutra and the Kauuika sutra. These serve the same purpose as the vidhana of the Rigveda and are of greater value in studying the Puranic-Vedic link than the AV text itself. There are several Upanishads that are associated with the AV, but appear to be relatively late additions to the tradition. The most important amongst these are the munaka and the pracna Upanishads.

***Issues of Note***

- The AV is the first Indic text dealing with medicine. It identifies the causes of disease as living causative agents such as the yatudhanya, the kimidi, the krimi and the durgama. The atharvans seek to kill them with a variety of drugs in order to counter the disease (see XIX.34.9). This approach to disease is surprisingly advanced compared to the trihumoral theory developed in the puragic era. Remnants of the original atharvanic thought did persist in the puragic era as can be seen in suruta's medical. Here following the atharvan theory the puragic text suggests germs as a cause for leprosy. In the same chapter sucruta also expands on the role of helminths in disease. These two can be directly traced back to the AV sachita. The hymn AV I.23-24 describes the disease leprosy and recommends the rajani aucadhi for its treatment. From the description of the aucadhi as black branching entity with dusky patches, it is very likely that is a lichen with antibiotic properties. Thus the AV can stake a claim to being one of the earliest texts to record uses of the antibiotic agents.

- The AV also informs us about warfare. A variety of devices such as an arrow with a duct for poison (apaskambha) and castor bean poison, poisoned net and hook traps, use of disease spreading bugs and smoke screens find a place in the AV sachita (eg. hymns IX.9, IX.10, the trioeacdi and nyarbudi hymns). These references to military practices and associated koeatriya rites were what gave the AV its formidable reputation. In the Mahabharata era that shortly followed after the end of the atharvan period there is a frequent comparison between weapons and the mantras of the heroes. Probably, this comparison was initially supposed to mean the application of deadly weapons as mentioned in the atharvan tradition.
- Several regular and special rituals of the Aryans are a major concern of the AV, just as in the three other vedas. The major rituals covered by the AV are marriage in kaC a-XIV and the funeral in kaC a-XVIII. There are also hymns that are specific to rituals of the bhigu-agnirasas, vatyas and khatriyas. One of the most important of these rites is the Viuhasahi Vrata, performed to invoke the icdra and Vicgu with the mantras of the XVIIth kaC a. The Vatya rituals were performed by individuals who took on a nomadic ascetic way of living and were generally sent into neighbouring states by the ruler of a particular state. They appear to have served a role in reconnaissance and negotiations with neighbouring states (compare with Arjuna's Vatya-like journey into the Yadu principality to woo Subhadra). Finally, there are some rituals aimed at the destruction of the enemies (Abhicharika hymns and rites) particularly using the closing mantras of the XVIth kaC a. While these support traditional negative views on the AV, in content they are mirrored by several other hymns from the Rig as well as the Yajuces. Moreover, Abhicharika rites were an integral part of

the vedic as amply attested in the brahmaga literature (see the tale of Yavakrd a in the Jaiminiya brahmaga). Thus the AV as such began fully within the classic Vedic fold, though it was more specific to certain clans of fire priests. The development of the Abhicharika rites to their more 'modern' form was seen only in the vidhana literature and in fact began within the Rigvedic tradition in the form of the Vigvidhana. The author of the (igvidhana provides passing reference to the development of similar rites in the AV tradition (the references to the Agnirasa Krityas). These rites reached their culmination in the Kauhika and Vaitana Sutra and in some of the Pariihthas (appendices) of the atharvan literature. However, these are far removed from the actual hymns themselves suggesting that they represent an encrustation on the atharvanic practice rather than its original form. While in its most extreme form Atharvanic Abhicharika faded away, it did seed the mainstream Hindu culture resulting in the origin of the Puranic form of the fire ritual (yaga-s). It also provided the launching pad for the worship of late evolving popular deities like Kumara and Ganapati to capture the mainstream Hindu ritual.

- *Philosophical Excursions:* The AV made the most important contributions to Aryan philosophical thought of all the Sachitas. One of the most spectacular expressions of this is seen in the hymn XII.I, the Hymn to goddess Earth or the Prithvi Suktam used in the Aghrayana rite. The foundations of Vaiceuika, the highest of the Hindu Darœanas is expressed in the mantra XII.1.26 in which the atoms (Pamsu) are described forming the stone, the stones agglutinating to form the rocks and the rocks held together to form the Earth. An early pantheistic thought (somewhat convergent to the latter day Vicihthadvaitins) is seen in the hymn X.7 that describes the common thread running through all manifest and un-manifest existence as the skaCbha. This skaCbha is described as what poured out of the Hiranya Garbha, that was the precursor of the complex world in a very simple form (X.7.28). (*Hiranya Garba* = "The radiant or golden egg or womb. Esoterically the luminious 'fire mist' or ethereal stuff from which the Universe was formed.")

This Skambha is Indra and Indra is the Skambha which describes all existence. The hymn also describes a pantheistic nature of the Vedic gods (X.7.38): skaCbha is the heat (tapa) that spreads through the universe (Bhuvana) as waves of water; the units of this spreading entity are the gods even as branches of one tree. This one theme that repeatedly presents itself in various interpretations that abounded in later Hindu philosophies and can be considered one of the most fundamental expression of Vedic thought.

***Dating:*** From alleged internal astronomical references (AVS XI.7), it has been surmised that the Atharvanic period included the time when the Pleiades occupied the spring equinox (roughly 2200 BC). Further, tradition suggests that paippalada, one of the early collators, and Vaidharbhi, one of the late contributors associated with the Atharvanic text, lived during the reign of prince Hiranyanabha of the Ikshvaku dynasty, interpreted to mean that the core AV composition was at least complete by 1500 BC.

While these approaches are not widely accepted as valid, it is clear that the core text of the AV is not particularly recent in the Vedic Sachita tradition, and falls within the classical *Mantra period* of Vedic Sanskrit in the late 2nd millennium BC—roughly contemporary with the Yajurveda mantras, the Rigvedic Khilani, and the redaction of the Samaveda.

The Atharvaveda is also the first Indic text to mention Iron (as uyama ayas, literally "black metal"), so that scholarly consensus dates the bulk of the Atharvaveda hymns to the early Indian Iron Age, corresponding to the 12th to 10th centuries BC or the early Kuru kingdom.

During its oral tradition, however, the text has been corrupted by later additions considerably more than the other Vedas, and it is only from comparative philology of the two surviving recensions that we may hope to arrive at an approximation of the original reading.

***Editions:*** The Shaunakiya text was edited 1960–62 by Vishva Bandhu, Hoshiarpur.

The bulk of the paippalada text was edited by Leray Carr Barret from 1905 to 1940 (book 6 by Edgerton, 1915) from a single Kashmirian uarada manuscript (now in Tübingen). This edition is outdated, since various other manuscripts were discovered in Bihar, Bengal and Orissa since. Some manuscripts are in the Orissa State Museum, but many manuscripts are in private possession, and are kept hidden by their owners. Many manuscripts were collected by Prof. Durgamohan Bhattacharya of Bengal by deceiving their owners, as told by his son Dipak Bhattacharya in 1968 (below), who describes the theft as valiant daredevilry:

*"The knowledge of the villagers, in whose possession many important manuscripts remain, about their possession is often very hazy [...] Prof. Bhattacharya secured a manuscript from an illiterate Brahmin on promise of return..."* (see: Zehnder (1999), p.19)

Books 1–15 were edited by Durgamohan Bhattacharya (1997). There is a provisional edition of book 20 by Dipak Bhattacharya. Book 2 and 5 were edited and translated by Thomas Zehnder (1999) and Alexander Lubotsky (2002), respectively.

## Samhita

The *Samhita* (Sanskrit: "joined" or "collected") is the basic text of each of the *Vedas*, comprising collections of hymns and ritual texts. This term was originally used in reference to the style of recitation used during hymns and chants. In the Vedas the Samhitas are supplemented by later explanatory commentaries, notably the Brahmanas and Upanishads. The term is also sometimes spelled 'Sanhita'.

***Other Samhitas:*** In later times, the term Samhita also came to be applied to the names of texts belonging to various fields of knowledge, not directly connected to the Vedas. *e.g.*

- Gherand Samhita
- Charak Samhita
- Ashtavakra Samhita
- Yogayajnavalkya samhita
- Brahma Samhita

## Gherand Samhita

*Gheranda Samhita* is one of the three classic texts of Hatha Yoga (the other two being the Hatha Yoga Pradapika and the Shiva Samhita). It is a late 17th Century text and is considered to be the most encyclopedic of the three classic texts on Hatha Yoga. Gheranda Samhita is a manual of Yoga taught by Gheranda to Chanda.

Unlike other Hatha Yoga texts, the Gheranda Samhita focuses upon the *Shat Kriyas* or internal body cleansing, *i.e.*, *Ghatastha Yoga*. The closing stanzas of the Gheranda Samhita deal with Samadhi, but the methods taught are different from Patanjali's methods.

## Charaka

*Charaka,* sometimes spelled *Caraka*, (300 BC) is one of the founders of Ayurveda. According to Charaka, health and disease are not predetermined and life may be prolonged by human effort.

***Contributions:*** According to the Charaka tradition, there existed six schools of medicine, founded by the disciples of the sage Punarvasu Atreya. Each of his disciples, *Agnivesha*, *Bhela*, *Jatukarna*, *Parashara*, *Harita*, and *Ksharapani*, composed a Samhita. Of these, the one composed by Agnivesha was considered the best. The Agnivesha Samhita was later revised by Charaka and it came to be known as Charaka Samhita. The Charaka Samhita was revised by Dridhbala. Charaka was from India. Ayurveda is traditionally divided into eight branches which, in Charaka's scheme, are:

1. Sutra-Sthana, general principles
2. Nidana-Sthana, pathology

3. Vimana-Sthana, diagnostics
4. Sharira-Sthana, physiology and anatomy
5. Indriya-Sthana, prognosis
6. Chikitsa-Sthana, therapeutics
7. Kalpa-Sthana, pharmacy
8. Siddhi-Sthana, successful treatment

***Charak Samhita:*** The Charak Samhita contains 120 adhyayas (chapters), divided into 8 parts.

1. Sutra Sthana (30 chapters)
2. Nidan Sthana (8 chapters)
3. Viman Sthana (8 chapters)
4. Sharir Sthana (8 chapters)
5. Indriya Sthana (12 chapters)
6. Chikitsa Sthana (30 chapters)
7. Kalpa Sthana (12 chapters)
8. Siddhi Sthana (12 chapters)

## Ashtavakra Gita

The *Ashtavakra Gita* (Song of Ashtavakra), also known by the name *Ashtavakra Samhita* is an influential non-dualist Vedic scripture traditionally said to have been spoken by the Sage Ashtavakra, though its authorship is not known with certainty.

There is little doubt though that it is very old, probably dating back to the days of the classic Vedanta period. The Sanskrit style and the doctrine expressed would seem to warrant this assessment. The work was known, appreciated and quoted by Ramakrishna and his disciple Vivekananda, as well as by Ramana Maharishi. Radhakrishnan refers to it with great respect, while even Sri Ravi Shankar has given a commentary on the scripture. Apart from that, the work speaks for itself. It presents the traditional teachings of Advaita Vedanta with a clarity and power very rarely matched.

The *Ashtavakra Gita* does not date itself or brand itself to any region, culture, or peoples, although it does contain Hindu-specific references, especially in the final chapter entitled 'I am Shiva' where it is proclaimed in the final verse that the author is in truth one with God. The essence of the *Ashtavakra Gita* is that there is no such thing as existence or non existence, right or wrong, or moral or immoral. In the eyes of the sage Ashtavakra who is the ostensible author of this text, one's true identity can be found by simply recognizing oneself as Pure Existence and that as individuals we are the awareness of all things.

The *Ashtavakra Gita* teaches that one is already free once one realises they are free. It advocates non-action (similar to the Daoist concept of Wu Wei), the loss of desire and severing of worldly attachments. To free oneself from the cycle of life and death, one should withdraw from all Earthly desires, worries and cares. To continue induldging in Earthly things even after one has realised their true nature is said to be foolish and time wasting. Instead it paints a picture of The Master as someone who continues to keep up their responsibilities in the world, not because they believe they have to or due to any worldy attachments, but simply that it is in their nature to do so.

It is important to avoid misinterpretation in this regard, and to that end teachers traditionally recommend that *Ashtavakra Gita* be pursued by only those who have already advanced on the spiritual path. Ashtavakra teaches that emotion, thought and even meditation can lead to further bondage and that religious practice itself is of no meaning to the enlightened.

## Brahmana

The *Brahmaas* are part of the Hindu *Shruti*; They are composed in Vedic Sanskrit, and the period of their composition is sometimes referred to as the *Brahmanic period* or *age* (approximately between 900 BC and 500 BC). They are essentially commentaries of the Vedas, explaining Vedic ritual. The earliest Brahmanas may have been written several centuries earlier, contemporary to the Black Yajurveda commentary prose, but they have only survived in fragments.

Each Brahmana is associated with one of the four Vedas, and within the tradition of that Veda with a particular shakha or school:

- Rigveda
  - o Shakala shakha: Aitareya Brahmana (AB)
  - o Bashkala shakha: Kaushitaki Brahmana (KS)
- Samaveda
  - o Kauthuma: PB, SadvB
  - o Jayminiya: Jayminiya Brahmana (JB)
- Yajurveda
  - o Krishna: the Brahmanas are integrated into the samhitas:
    - * Maitrayani (MS)
    - * Carakakatha (CS)
    - * Kapisthalakatha (KS)
    - * Taittiriya (TS). The Taittiriya school has an additional Taittiriya Brahmana (TB)
  - o Shukla
    - * Vajasaneyi Madhyandina: Shatapatha Brahmana, Madhyadina recession (ShB)
    - * Kanva: Shatapatha Brahmana, Kanva recession (ShBK)
- Atharvaveda
  - o Paippalada: Gopatha Brahmana

***Aitareya Brahamana***

*Aitareya Brahmana*, also known as the *Ashvalayana Brahmana* is associated to Rigveda, first among the four Vedas. It is believed to have been composed around 600 BC and is perhaps the oldest Brahmana. It deals principally with Soma sacrifices.

***Jaiminiya Upanishad Brahmana***

The *Jaiminiya Upanishad Brahmana* (JUB) is a Vedic text associated with the Jaiminiya shakha of the Samaveda. It may be considered a very early Upanishad, together with the

Brhadaragyaka and Chandogya Upanishads dating to the Brahmana period of Vedic Sanskrit, likely predating the 6th century BC.

It is not to be confused with the Jaiminiya Brahmana (JB), the actual Brahmana commentary of the Jaiminiya school.

### *Kaushitaki Brahmana*

The *Kaushitaki Brahmana* is the Brahmana associated with the Rigveda in the Bashkala shakha.

### *Panchavimsha Brahmana*

The *Tandhya-maha-*(or *Praudha-*) *brahmana*, or "great" Brahmana — usually called *Panchavimsha-brahmana* from its consisting of twenty-five *adhyayas* is a Brahmana of the Samaveda, treating of the duties of the udgatars generally, and especially of the various kinds of chants.

### *Shatapatha Brahmana*

*Shatapatha Brahmana* is one of the prose texts describing the Vedic ritual, associated with the White Yajurveda. It survives in two recensions, *Madhyandina* (*CEBM*, of the *vajasaneyi madhyandina* shakha) and *Kanva* (*CEBK*, of the *kagva* shakha), with the former having the eponymous 100 brahmanas in 14 books, and the latter 104 brahmanas in 17 books.

The CEB is notable as one of the oldest prose (non-metrical) Sanskrit texts altogether. Linguistically, it belongs to the Brahmana period of Vedic Sanskrit, dated to the first half of the 1st millennium BC (roughly 800 BC).

Among the points of interest are the mythological sections embedded in it, including myths of creation and the Deluge of Manu. The text describes in great detail the preparation of altars, ceremonial objects, ritual recitations, and the Soma libation, along with the symbolic attributes of every aspect of the rituals.

Some Hindu scholars have dated it to around 1800 BC, based on the reference in it of migration from the Saraswati river area to east India, because the river is said to have dried up around 1900 BC. Archaeoastronomers have dated it to around

2000 BC based on a reference to the Pleiades (Krttikas) "rising in the east" at autumnal equinox (CEBM 2.1.2.1). Such dating interpretations are controversial. (see Hindu astronomy)

The 14 books of the Madhyandina recession can be divided into two major parts. The first 9 books have close textual commentaries, often line by line, of the first 18 books of the corresponding samhita of the Yajurveda. The following 5 books cover supplementary and ritualistically newer material, besides including the celebrated Brhadaranyaka Upanishad as most of the 14th and last book.

### *Taittiriya Brahmana*

The *Taittiriya Brahmana* (TB) is the Brahmana associated with the Taittiriya shakha of the Black Yajurveda. The Taittiriya school is the only school of the Black Yajurveda to have a Brahmana additional to the commentary already interspersed in the Samhita.

## Aranyaka

The *Aranyakas* are part of the Hindu Kruti; these religious scriptures are sometimes argued to be part of either the Brahmanas or Upanishads. The name translates to "the forest books", meaning, treatises for hermits or sadhus living in the wilderness. This contrasts with the grhyasutras, treatises intended for domestic life. Their language is early Classical Sanskrit, and together with the bulk of the Upanishads, the Aranyakas form the basis of Vedanta, roughly dating to a few centuries before the Common Era.

***Books:*** The Aranyakas discuss philosophy and sacrifice. They are believed to have originated with the various mystical ascetic groups that developed in post-Vedic India. The Aranyakas constitute a more philosophical and mystical interpretation of the themes presented in the Vedas, as opposed to the Brahmanas, which were primarily concerned with the proper performance of ritual. Like the Upanishads, the Aranyakas may have initially constituted a secret or hidden teaching, not in the sense of being forbidden or restricted, but rather being both a non-obvious expansion on the themes of the Vedas and a teaching that was primarily conveyed individually from teacher to student.

The Aranyakas are associated with and named after individual Vedic shakhas.

- Aitareya Aranyaka belongs to the Shakala Shakha of Rigveda
- Taittiriya Aranyaka belongs to the Taïttiriya Shakha of Krishna-Yajurveda
- Katha Aranyaka belongs to the Kaṭha-Charaka Shakha of the Krishna-Yajurveda
- Kaushitaki Aranyaka belongs to the Kaushitaki and Shankhayana Shakhas of Rigveda
- Maitrayaniya Aranyaka belongs to the Maitrayaniya Shakha of Krishna-Yajurveda
- Talavakara Aranyaka belongs to the Talavakara or Jaiminiya Shakha of Samaveda

The Atharvaveda has no surviving Aranyaka, although indications are that there did exist Araynaka works attached to this Veda in the past.

***The Secret of the Brahmanas:*** There is also a certain continuity of the Aranyakas from the Brahmanas in the sense that the Aranyakas go into the subtle esoteric meanings of the ritualistic tilt that the rites detailed in the Brahmanas give to the mantras of the Vedas. It is this leap into subtlety that provides the reason for Durgacharya in his commentary on the Niruktas to say that the Aranyakas are 'Rahasya Brahmana', that is, the Secret of the Brahmanas. In short, the undercurrent of emphasis of the Aranyakas is to point out that through all the different multiplicities that one has to contend with, there is a single thread of continuity, namely, the One Absolute Reality. The language of the Aranyakas is, unlike Vedic Sanskrit, nearer to the ordinary Sanskrit of the layman.

A glimpse into the contents of some of the Aranyakas

***Aitareya Aranyaka:*** There are five chapters each of which is even considered as a full Aranyaka. The first one deals with the regimen known as 'Mahaa-vrata'. The explanations are both ritualistic as well as absolutistic. The second one has six chapters of which the first three are about 'Praana-vidyaa' – meaning, Prana, the Vital Air that constitutes the life-breath of a living body is also the life-breath of all mantras, all vedas and all vedic declarations (cf. 2.2.2 of aitareya Aranyaka). It is in this portion of the Aranyaka that one finds specific statements about how one who follows the vedic injunctions and performs the sacrifices goes to become the God of Fire, or the Sun or Air and how one who transgresses the vedic prescriptions is born into lower levels of being, namely, as birds and reptiles.

It is in Aitareya Aranyaka Praana is recommended to be worshipped in the form of Rishis. Praana is Vishvamitra, because all the universe ('vishwa') is the object of experience of this praana deity. Praana is Vamadeva, because the word "Vaama" indicates respectability and deservingness to be worshipped and served. It is also Atri Maharishi because, the word 'traayate' in Atri indicates the Praana that protects from sins. The Praana enters this body and supports it, therefore it is Bharadvaja – where the 'bhara' word denotes supporting and the 'Vaaja' indicates the mortal body that is made mobile by the residence of praaana. Vashishhta is also Praana because the word 'vas' indicates the dwelling in this body of the senses made possible by praana.

The 4th, 5th and 6th chapters of this second Aranyaka constitute what is known as Aitareya Upanishad.

The third Aranyaka in this chain of Aranyakas is also known as 'Samhitopanishad'. This elaborates on the various ways – like pada-paatha, krama-paatha, etc.—of reciting the Vedas and the nuances of the 'svaras'.

The fourth and the fifth Aranyakas are technical and dwell respectively on the mantras known as 'MahaanaamnI' and the yajna known as 'Madhyandina'.

***Shankhayana Aranyaka:*** There are fifteen chapters here. From the third to the sixth it constitutes the Kaushitaki Upanishad. The seventh and eighth are known as a Samhitopanishad.

The first two chapters deal with the Mahavrata. The ninth talks about the greatness of Prana. The tenth chapter deals with the esoteric implications of the agnihotra ritual. All divine personalities are inherent in the Purusha, just as Agni in speech, Vayu in Prana, the Sun in the eyes, the Moon in the mind, the directions in the ears and water in the potency. The one who knows this, says the Aranyaka, and in the strength of that conviction goes about eating, walking, taking and giving, satisfies all the gods and what he offers in the fire reaches those gods in heaven. (cf.10-1).

The eleventh chapter prescribes several antidotes in the form of rituals for warding off death and sickness. It also details the effects of dreams. The 12th chapter elaborates the fruits of prayer. The 13th gets into more philosophical matters and says one must first attitudinally discard one's bodily attachment and then carry on the 'shravana', manana and nidhidhyasana and practise all the disciplines of penance, faith, self-control, etc.

The 14th gives just two mantras. One extols the "I am Brahman" mantra and says it is the apex of all Vedic mantras. The second mantra declares that one who does not get the meaning of mantras but only recites vedic chants is like an animal which does not know the value of the weight it carries. The final chapter gives a long list of genealogy of spiritual teachers from Brahma downwards upto Guna-Sankhayana.

***Brihad-Aranyaka:*** This is the famous Upanishad of that name. The Self is the subject of discussion here from all aspects. For a complete discussion see Brihad-Aranyaka Upanishad.

***Taittiriya-Aranyaka:*** There are ten chapters here. The 7th, 8th and 9th constitute the well-known Taittiriya Upanishad. The tenth is a long Upanishad known as Maha-Narayana-Upanishad; it contains several important mantras culled from the three vedas. Chapters one to six form the Aranyaka proper.

The first one is the famous Surya namaskara chapter. The second one is a description of the five maha-yajnas that every Brahmin has to do daily. Naturally the sacred thread, the yajnopavita, of the Brahmin is extolled and elaborated here.

The sandhya worship, the worship of the manes, worship of the brahman through the brahma-yajna, the cleansing homa-sacrifice called the kushmanda-homa are all dealt with in detail. In this chapter the word 'shramana' is used (2-7-1) in the meaning of a doer of penance (tapasvii); this word came to mean in later times, a recluse of the Buddhist and Jain religions.

The third and fourth chapters go into further technicalities of several other homas and yajnas. The fourth chapter has also sections on mantras that may be used for averting (or causing !) havoc. The fifth is an academic treatise on yajnas. The sixth one is a collection of 'pitr-medha' mantras, that is, the mantras recited on the occasion of, and used for, the rituals for the disposal of the dead body.

## Upanishad

The *Upanishads* are part of the Vedas and form the Hindu scriptures which primarily discuss philosophy, meditation and nature of God; they form the core spiritual thought of Vedantic Hinduism. The Upanishads are mystic or spiritual contemplations of the Vedas, their putative end and essence, and thus known as *Vedanta* ("the end/culmination of the Vedas"). The Upanishads were composed over several centuries. The oldest, such as the Brhadaranyaka and Chandogya Upanishads, have been dated to around the eighth century BCE. The roots of many Indian religions are built upon the foundation of the Upanishads.

***Etymology:*** The Sanskrit term *upanicad* derives from *upa-* (near), *ni-*(down) and *sad* (to sit), *i.e.* referring to the "sitting down near" a spiritual teacher (guru) in order to receive instruction in the Guru-shishya tradition or parampara. The teachers and students appear in a variety of settings (husband answering questions about immortality, a teenage boy being taught by Yama, etc.).

***Traditional Etymology***

*Upa + ni + cad (sad)*

*Upa* = an *upasarga* (prefix) meaning *Samipyam*-Nearest, *i.e. Atma* (Oneself);

*ni* = an *upasarga* (prefix) meaning *Niœcaya*-definitely ascertained knowledge;

*cad* = a *dhatu* (verbal root) which was originally *sad*. It becomes *cad* due to the affix *kvip*, which forms a noun in the sense of the verbal action. *Sad* has three meanings: (1) *Viharagam*-dissolves, disintegrates, destroys, (2) *Gati* or *Prapti*-go, lead, attain, know, (3) *Avasadana*—puts an end to.

Thus, *Upanicad* means:

- Definitely ascertained knowledge of *Atma* (Oneself) which dissolves, disintegrates, destroys ignorance.
- Definitely ascertained knowledge of *Atma* (Oneself) which leads one to *Brahman*; makes one gain, know *Brahman*.
- Definitely ascertained knowledge of *Atma* (Oneself) which puts an end to *sacsara* (a life of becoming, characterized by birth, death, suffering, etc.)

Definition of *Upanicad* based on the Introductory *Uagkarabhacyam* on *Kamha Upanicad* and *Brhadaragyaka Upanicad*. The word *Upanicad* ultimately means Self-knowledge = *Atmavidya*, Knowledge of *Brahman* = *Brahmavidya*. The *Magukya Upanicad Mahavakya* (verse 2) *Ayam Atma Brahma*. "This *Atma* is *Brahman*," confirms the view of the *Upanicad*, that in truth, you (the individual self, free from incidental attributes) are *Brahman* (the infinite and the ultimate basis of all reality).

***Major Upanishads:*** Different Upanishads are affiliated with the four Vedas (Rigveda, Yajurveda, Samaveda and Atharvaveda). The Upanishads were transmitted orally by the Vedic school sakhas. The longest and oldest Upanishad, are the Brhadaragyaka and the Chandogya respectively.

The language of the Upanishads, is Sanskrit, the oldest among them still classifying as late Vedic Sanskrit. The oldest Upanishads, the Brhadaragyaka and the Chandogya are composed in prose. These early texts may date back to the 8th-7th centuries BCE. Later followed a series of Upanishads composed in verse, such as the Ica, Magd.ukya, Katha, and bvetacvatara Upanishads.

According to tradition, there were over two hundred *Upanishads*, but the philosopher and commentator Shankara only composed commentaries to eleven of them. The Upanishads commented on by Shankara are generally regarded as the oldest ones. The Muktika Upanishad lists 108 Upanishads. In 1656, at the order of Dara Shikoh, the Upanishads were translated from Sanskrit into Persian.

These philosophical and meditative tracts form the backbone of Hindu thought. Of the early Upanishads, the Aitareya and Kauoitaki belong to the Rigveda, Kena and Chandogya to the Samaveda, Ioa and Taittiriya and Bhadara yaka to the Yajurveda, and Pracna and Mudaka to the Atharvaveda. In addition, the Madukya, Katho, 'veta'vatara are very important. Others also include Mahanarayaa and Maitreyi Upanishads as key.

***Place in the Hindu Canon:*** Scholars of the Vedic books consider the four Vedas as poetic liturgy, collectively called *mantra* or *samhita*—adoration and supplication to the deities of Vedic religion, in parts already melded with monist and henotheist notions, and an overarching order (Rta) that transcended even the Gods.

The *Brahmana* were a collection of ritual instructions, books detailing the priestly functions (which first were available to all men, and so concretized into strictly Brahmin privilege). These came after the Mantra.

Vedanta, is chiefly composed of *Aranyakas* and Upanishads. The Aranyakas ("of the forest") detail meditative yogic practices, contemplations of the mystic one and the manifold manifested principles. The Upanishad basically realized all the monist and universal mystical ideas that started in earlier Vedic hymns, and have exerted an influence unprecedented on the rest of Hindu and Indian philosophy. However, by adherents they are not considered philosophy alone, and form meditations and practical teachings for those advanced enough to benefit from their wisdom.

***Contents:*** The *Taittiriya Upanishad* says this in the Ninth Chapter:

*He who knows the Bliss of Brahman, whence words together with the mind turn away, unable to reach It? He is not afraid of anything whatsoever. He does not distress himself with the thought: "Why did I not do what is good? Why did I do what is evil?". Whosoever knows this regards both these as Atman; indeed he cherishes both these as Atman. Such, indeed, is the Upanishad, the secret knowledge of Brahman.*

***Taittiriya Upanishad Capt. 9 (II-9-1):*** The Upanishads hold information on basic Hindu beliefs, including belief in a world soul, a universal spirit, Brahman, and an individual soul, Atman (Smith 10). A variety of lesser gods are seen as aspects of this one divine ground, Brahman (different from Brahma). Brahman is the ultimate, both transcendent and immanent, the absolute infinite existence, the sum total of all that ever is, was, or ever shall be.

For Advaita (non-dual) philosophers Brahman is not a God in the monotheistic sense, as they do not ascribe to it any limiting characteristics, not even those of being and non-being, and this is reflected in the fact that in Sanskrit, the word Brahman has no gender (masculine or feminine or neuter). Dvaita philosophy holds that Brahman is ultimately a personal God, Vishnu, or Krishna (*brahmano hi pratisthaham*, Bhagavad Gita 14.27).

"Who is the Knower?" "What makes my mind think?" "Does life have a purpose, or is it governed by chance?" "What is the cause of the Cosmos?"

The sages of the Upanishad try to solve these mysteries and seek knowledge of a Reality beyond ordinary knowing. They also show a preoccupation with states of consciousness, and observed and analysed dreams as well as dreamless sleep.

***Philosophy:*** Due to their mystical nature and intense philosophical bent that does away with all ritual and completely embraces principals of One Brahman and the inner Atman (Self), the Upanishads have a universal feel that has led to their explication in numerous manners, giving birth to the three schools of Vedanta.

The Vedantin philosopher Adi Shankara summed up all the Upanishad in one phrase and said that in the end, the ultimate, formless, inconceivable Brahman is the same as our soul, Atman. We only have to realize it through discrimination and piercing through Maya.

A distinctive quotation that is indicative of the call to self-realization, one that inspired Somerset Maugham in titling a book he wrote on Christopher Isherwood, is as follows:

*Get up! Wake up! Seek the guidance of an Illumined teacher and realize the Self. Sharp like a razor's edge is the path, The sages say, difficult to traverse.*

***Death Instructing Nachiketa in the Katho (Word) Upanishad:*** The Upanishads also contain the first and most definitive explications of 'aum' as the divine word, the cosmic vibration that underlies all existence and contains multiple trinities of being and principles subsumed into its One Self. The *Isha* says of the Self (Verses 6, 7 & 8 of Isha Upanishad):

> *Whoever sees all beings in the soul and the soul in all beings does not shrink away from this. In whom all beings have become one with the knowing soul what delusion or sorrow is there for the one who sees unity? It has filled all. It is radiant, incorporeal, invulnerable, without tendons, pure, untouched by evil. Wise, intelligent, encompassing, self-existent, it organizes objects throughout eternity.*

*Isha Upanishad Verses 6, 7, & 8: "Aum Shanti Shanti Shanti"* This, too, is found first in the Upanishads, the call for tranquillity, for divine stillness, for Peace everlasting.

Dara Shikoh, the Muslim sufi, and son of Mughal emperor Shah Jahan, translated the Upanishads in Persian in order to find in it elements of monotheism that might pave the way for a common mystical bond between Islam and Hinduism.

## *List of Upanishads*

***"Principal" Upanishads:*** The following is a list of the eleven "principal" (*mukhya*) Upanishads that were commented upon by Shankara, and that are accepted as shruti by all

Hindus. They are listed with their associated Veda (Rigveda (ZV), Samaveda (SV), White Yajurveda (YV), Black Yajurveda (KYV), Atharvaveda (AV)).

1. Aitareya (ZV)
2. Brhadaragyaka (YV)
3. Ica (YV)
4. Taittiriya (KYV)
5. Kamha (KYV)
6. Chandogya (SV)
7. Kena (SV)
8. Muaka (AV)
9. Maukya (AV)
10. Praœna (AV)
11. Uvetauvatara

The Kauuitaki and Maitrayagi Upanishads are sometimes added to extend the canon to 13. They are also the oldest Upanishads, likely all of them dating to before the Common Era. From linguistic evidence, the oldest among them are likely the Brhadaragyaka and Chandogya Upanishads, belonging to the late Vedic Sanskrit period; the remaining ones are at the transition from Vedic to Classical Sanskrit.

***Canon by Vedic Shakha:*** The older Upanishads are associated with Vedic Charanas (Shakhas or schools).

The Aitareya Upanishad with the Shakala shakha, the Kauuitaki Upanishad with the Bashakala shakha; the Chandogya Upanishad with the Kauthuma shakha, the Kena Upanishad, and the Jaiminiya Upanishad Brahmana, with the Jaiminiya shakha; the Kamha Upanishad with the Caraka-Katha shakha, the Taittiriya and Uvetauvatara with the Taittiriya shakha; the Maitrayagi Upanishad with the Maitrayani shakha; the Brhadaragyaka and Ica Upanishads with the Vajasaneyi Madhyandina shakha, and the Maukya and Muaka Upanishads with the Shaunaka shakha. Additionally, parts of earlier texts, of Brahmanas or passages of the Vedas themselves, are sometimes considered Upanishads.

***The Muktika Canon:*** The following is a list of the 108 canonical Upanishads of the Advaita school, according to the Muktika Upanishad (number 108), 1:30-39 (which does not list the associated Veda). In this canon,

- 10 Upanishads are associated with the Rigveda and have the Shanti beginning *vagme-manasi*.
- 16 Upanishads are associated with the Samaveda and have the Shanti beginning *apyayantu*.
- 19 Upanishads are associated with the White Yajurveda and have the Shanti beginning *purgamada*.
- 32 Upanishads are associated with the Black Yajurveda and have the Shanti beginning *sahanavavatu*.
- 31 Upanishads are associated with the Atharvaveda and have the Shanti beginning *bhadram-kargebhi*.

The first 10 are grouped as *mukhya* "principal", and are identical to those listed above. 21 are grouped as Samanya Vedanta "common Vedanta", 23 as Sannyasa, 9 as Shakta, 13 as Vaishnava, 14 as Shaiva and 17 as Yoga Upanishads.

1. Ica, (YV, Mukhya) "The Inner Ruler"
2. Kena (SV, Mukhya) "Who moves the world?"
3. Kamha (KYV, Mukhya) "Death as Teacher"
4. Prakna, (AV, Mukhya) "The Breath of Life"
5. Muaka (AV, Mukhya) "Two modes of Knowing"
6. Maukya (AV, Mukhya) "Consciousness and its phases"
7. Taittiriya (KYV, Mukhya) "From Food to Joy"
8. Aitareya, (ZV Mukhya) "The Microcosm of Man"
9. Chandogya (SV, Mukhya) "Song and Sacrifice"
10. Brhadaragyaka (YV, Mukhya)
11. Brahma (KYV, Sannyasa)
12. Kaivalya (KYV, Shaiva)
13. Jabala (YV, Sannyasa)
14. Uvetauvatara (KYV, Sannyasa) "The Faces of God"
15. HaCsa (YV, Yoga)
16. AruGeya (SV, Sannyasa)

17. Garbha (KYV, Sannyasa)
18. NarayaGa (KYV, Vaishnava)
19. Paramahacsa (KYV, Sannyasa)
20. Am[tabindu (KYV, Yoga)
21. Amrtanada (KYV, Yoga)
22. Kira (AV, Shaiva)
23. Atharvakikha (AV, Shaiva)
24. Maitrayagi (SV, Sannyasa)
25. Kauuitaki (ZV, Samanya)
26. Brhajjabala (AV, Shaiva)
27. Nsichatapani (AV, Vaishnava)
28. Kalagnirudra (KYV, Shaiva)
29. Maitreyi (SV, Sannyasa)
30. Subala (KYV, Samanya)
31. Kcurika (KYV, Yoga)
32. Mantrika (KYV, Samanya)
33. Sarvasara (KYV, Samanya)
34. Niralamba (KYV, Samanya)
35. Kukarahasya (KYV, Samanya)
36. Vajrasuchi (SV, Samanya)
37. Tejobindu (KYV, Sannyasa)
38. Nadabindu (ZV, Yoga)
39. Dhyanabindu (KYV, Yoga)
40. Brahmavidya (KYV, Yoga)
41. Yogatattva (KYV, Yoga)
42. Atmabodha (ZV, Samanya)
43. Parivrat (Naradaparivrajaka) (AV, Sannyasa)
44. Trikikhi (KYV, Yoga)
45. Sita (AV, Shakta)
46. Yogachuamagi (SV, Yoga)
47. Nirvaga (ZV, Sannyasa)
48. Maalabrahmaga (KYV, Yoga)
49. Dakcigamurti (KYV, Shaiva)

50. Karabha (AV, Shaiva)
51. Skanda (Tripavibhumi) (KYV, Samanya)
52. Mahanarayaga (AV, Vaishnava)
53. Advayataraka (KYV, Sannyasa)
54. Ramarahasya (AV, Vaishnava)
55. Ramatapagi (AV, Vaishnava)
56. Vasudeva (SV, Vaishnava)
57. Mudgala (ZV, Samanya)
58. Uailya (AV, Yoga)
59. Paiegala (KYV, Samanya)
60. Bhikcu (KYV, Sannyasa)
61. Mahad (SV, Samanya)
62. Uariraka (KYV, Samanya)
63. Yogauikha (KYV Yoga)
64. Turiyatita (KYV, Sannyasa)
65. Sannyasa (SV, Sannyasa)
66. Paramahacsaparivrajaka (AV, Sannyasa)
67. Akcamalika (Malika) (ZV, Shaiva)
68. Avyakta (SV, Vaishnava)
69. Ekakcara (KYV, Samanya)
70. Annapurga (AV, Shakta)
71. Surya (AV, Samanya)
72. Akci (KYV, Samanya)
73. Adhyatma (KYV, Samanya)
74. Kuika (SV, Sannyasa)
75. Savitri (SV, Samanya)
76. Atma (AV, Samanya)
77. Pauupata (AV, Yoga)
78. Parabrahma (AV, Sannyasa)
79. Avadhuta (KYV, Sannyasa)
80. Devi (AV, Shakta)
81. Tripuratapani (AV, Shakta)
82. Tripura (ZV, Shakta)

83. Kamharudra (KYV, Sannyasa)
84. Bhavana (AV, Shakta)
85. Rudrahrdaya (KYV, Shaiva)
86. Yogakualini (KYV, Yoga)
87. Bhasma (AV, Shaiva)
88. Rudrakca (SV, Shaiva)
89. Gagapati (AV, Shaiva)
90. Darkana (SV, Yoga)
91. Tarasara (KYV, Vaishnava)
92. Mahavakya (AV, Yoga)
93. Pancabrahma (KYV, Shaiva)
94. Pragagnihotra (KYV, Samanya)
95. Gopalatapani (AV, Vaishnava)
96. Krcga (AV, Vaishnava)
97. Yajnavalkya (KYV, Sannyasa)
98. Varaha (KYV, Sannyasa)
99. Uamyayani (KYV, Sannyasa)
100. Hayagriva (AV, Vaishnava)
101. Dattatreya (AV, Vaishnava)
102. Gara (AV, Vaishnava)
103. Kali-Sagmaraga (Kali) (KYV, Vaishnava)
104. Jabala (SV, Shaiva)
105. Saubhagya (ZV, Shakta)
106. Sarasvatirahasya (KYV, Shakta)
107. Bahvrca (ZV, Shakta)
108. Muktika (KYV, Samanya)

## Aitareya Upanishad

The *Aitareya* Upanishad is one of the older, "primary" Upanishads commented upon by Shankara. It is a Mukhya Upanishad, associated with the Rigveda. It figures as number 8 in the Muktika canon of 108 Upanishads.

The Aitareya Upanishad is a short prose text in three chapters and contains 33 verses in total. It comprises 4-6 chapters of the second book of the older vedic text, Aitareya Aranyaka.

In the first chapter of the text, atman, the inner self, is portrayed as a divine creator. In the second chapter, the three births of the atman are described.

The third chapter deals with the qualities of the Self or Brahman. It contains one of the most famous expressions of the Vedanta, "pragnanam Brahma," which is one of the Mahavakyas.

## Brihadaranyaka Upanishad

The *Brihadaranyaka* Upanishad is one of the older, "primary" (*mukhya*) Upanishads commented upon by Adi Shankara. It is contained within the Shatapatha Brahmana, and its status as an independent Upanishad may be considered a secondary extraction of a portion of the Brahmana text. This makes it one of the oldest (if not the oldest) texts of the Upanishad corpus, possibly dating to as early as the 9th century BCE. It is associated with the White Yajurveda. It figures as number 10 in the Muktika canon of 108 Upanishads.

## Isha Upanishad

The *Isha Upanishad* (*iua upanicad*) or *Ishopanishad* (*iuopanicad*), also known as the *Ishavasya Upanishad* (*iuavasya upanicad*), is a Sanskrit poem (or sequence of mantras) from the Upanishads and is considered shruti by followers of a number of diverse traditions within Hinduism. It is one of the smaller texts of the Upanishads (18 verses in total) but also one of the most often quoted.

***Background and Content:*** The Isha Upanishad is contained in the final *adhyaya* (chapter) of the White Yajurveda (VS). It consists of either 17 or 18 two-line verses covering a wide spectrum of philosophy, religion, ritualism and metaphysics in a concise manner. The name of the Upanishad derives from its incipit, *Iuavasyam idam sarvam,* "This universe is enveloped by the Lord". According to one Western scholar, it belongs to an advanced stage of monist speculation, assuming a Lord (*ish*) of the universe (Weber 1878:103). One view has it that this is one of the later mukhya Upanishads, dating approximately to Mauryan times. According to Mahidhara's commentary, it is a polemic against the Bauddhas (the predecessors of the later Samkhya doctrines).

Mahatma Gandhi used to say, even if every other scripture of Hinduism vanished and this alone survived, Hinduism will survive.

***Text:*** In the two shakhas of the White Yajurveda, VSM and VSK, the order of verses 1-8 is the same. VSK verses 9–14 correspond to VSM verses 12, 13, 14, 9, 10, 11. VSM 17 is a variation of VSK 15, VSK 16 is lacking in VSM, and VSK 17–18 correspond to VSM 15–16. Verse numbers in this article refer to VSK:

| VSK | 1 | 2 | 3 | 4 | 5 | 6 | 7 | 8 | 9 | 10 | 11 | 12 | 13 | 14 | 15 | 16 | 17 | 18 |
|---|---|---|---|---|---|---|---|---|---|---|---|---|---|---|---|---|---|---|
| VSM | 1 | 2 | 3 | 4 | 5 | 6 | 7 | 8 | 12 | 13 | 14 | 9 | 10 | 11 | (17) | - | 15 | 16 |

***Verse 1***

*iuavasyam idam sarva? yat kinca jagatya? jagat*
*tena tyaktena bhunjitha ma grdha kasya svid dhanam*

Transliteration: idam sarvam = all this (this entire universe); Isavasyam = Isasya avasyam = Isasya avasayogyam = pervaded by Isa, the Lord Hari; yat kincha = also whatsoever, and whatever there may be; jagatyam = prakrtyam = in the primordial nature; jagat = the world; tena = by Him (the Lord Hari); tyaktena = dattena vittena = wealth granted, allotted, given; bhunjitha = bhogam kuryah = do experience, enjoy; ma grdhah = ma kanksethah = do not crave for, do not seek; kasyasvid = any one else's; dhanam = wealth.

*"Everything animate or inanimate that is within the universe is controlled and owned by the Lord. One should therefore accept only those things necessary for himself, which are set aside as his quota, and one should not accept other things, knowing well to whom they belong." (trans. Prabhupada)*

*"The world is swaddled in the glory of the Lord.*
*Renounce it and enjoy it. Do not covet anyone's wealth."*
*(trans. P. Lal)*

***Verse 7***

*"One who always sees all living entities as spiritual sparks, in quality one with the Lord, becomes a true knower of things. What, then, can be illusion or anxiety for him?" (trans. Prabhupada)*

***Verse 8***

*Sa paryagac chukram akayam avragam asnaviram uuddham apapa-viddham*
*Kavir manici paribhu svayambhur yathatathyato 'rthan vyadadhac chauvatibhya samabhya*

"He hath attained [*paryagat*] unto the Bright [*shukram*], Bodiless, Woundless, Sinewless, the Pure which evil hath not pierced."

"Far-sighted, wise, encompassing, he self-existent hath prescribed aims, as propriety demands, unto the everlasting Years" (trans. Griffith).

"Such a person must factually know [*paryagat*] the greatest of all, the Personality of Godhead [*shukram*], who is unembodied, omniscient, beyond reproach, without veins, pure and uncontaminated, the self-sufficient philosopher who has been fulfilling everyone's desire since time immemorial." (trans. Prabhupada). According to Shankara and Mahidhara, the neuter adjectives *shukram*, etc. relate to Atman (*sas* "He") rather than to Brahma or Highest Essence. Depending on the interpretation of the first line, "Far-sighted", etc. refers either to the man who knows, or to Atman itself.

***Verse 18***

"O my Lord, as powerful as fire, O omnipotent one, now I offer You all obeisances, falling on the ground at Your feet. O my Lord, please lead me on the right path to reach You, and since You know all that I have done in the past, please free me from the reactions to my past sins so that there will be no hindrance to my progress." trans. Prabhupada.

***The Supreme:*** The *Isha Upanishad* is significant amongst the Upanishads for its description of the nature of the Supreme Person as the controller (Ish) behind our universe. 'He who is unembodied, omniscient, beyond reproach, without veins, pure and uncontaminated' (verse 8). (As stated above, commentators differ as to whether the referrent here is Ish, Atman, or he who has recognized Atman.) He who 'walks and does not walk', who is 'far away, but very near as well', who 'although fixed in His abode is swifter than the mind' (from verses 4 & 5). In this way the Upanishad gives a number of seemingly paradoxical descriptions of the Supreme.

The later verses take the form of a series of prayers requesting that the speaker be able to see past the supreme light or effulgence in order to understand the true nature of the Supreme Lord Himself and be freed of the sins of past misdeeds

***Morality:*** The text also discusses the nature of karma for the living entities, and it promotes a way of life of simple living, 'accepting only those things set aside for oneself' (text 1) and a culture of knowledge over ignorance (text 10). The text also mentions negative reactions for one who kills another living being, and it advises seeing all living entities in connection with the Supreme Person and thus being equipoised towards them (verses 6 & 7).

Worship of devas (demigods) is condemned in favour of worship of the Supreme God only, for worship of anything other than the Supreme brings about different results. The reward for correct worship is given as a future life of eternity, bliss and knowledge.

## Taittiriya Upanishad

The *Taittiriya* Upanishad is one of the older, "primary" Upanishads commented upon by Shankara. It is associated with the *Taittiriya* school of the Black Yajurveda. It figures as number 7 in the Muktika canon of 108 Upanishads. Black Yajurveda is also called Taittiriya Samhita. Like all vedas Black Yajurveda consists of Samhita, Brahmana and Aranyaka parts. The eighth and ninth chapters of Taittiriya Aranyaka form the Taittiriya Upanishad. The tenth chapter is famous as Mahanarayana Upanishad or Yajnikee.

It is divided into three sections called Vallis. The *Siksha Valli*, the *Brahmananda Valli* and the *Bhrigu Valli*. For the convenience of study, each Valli is subdivided into smaller verses called Anuvak.

***Shiksha Valli:*** The Siksha Valli deals with the discipline of Shiksha (which is the first of the six Vedangas or "limbs" or auxiliaries of the Vedas). It is concerned with phonetics and pronunciation. Main intention of starting Upanishad with Shiksha seems to be that, though meaning of vedic Mantras is important, one should not give any less importance to the right pronouciation and phonetics of mantras. If Mantras are pronounced correctly then meaning becomes clear to recitor. Also there is a belief that Mantras uttered with incorrect pronouciation loose their power.

First Anuvak of this Valli starts with Shanti Mantra (Peace Prayer) "OM Sham no Mitra..". Purpose of having Shanti mantra in the beginning is to please the gods and remove any obstacles during the study of Upanishads.

Second Anuvak defines the contents of Shiksha discipline.

Third Anuvak tells about the intimate connection between the syllables of vedic verses or mantras using five examples and implicitly tells that one should meditate on those examples to realise connections told in them. Also declares that one who realised those connections will get good fruits including heaven.

Fourth Anuvak consists of Mantras and Homas to be done by those who aspire for devine knowledge (Medha kama) and wealth (Shree kama). In this anuvak teacher prays that "As water flows from high land to low land, as months join to become year, let good desciples come to me from everywhere"

Fifth and Sixth Anuvaks try to describe Brahman in the form of "Vyahruti"s. The words "Bhooh", "Bhuvah", "Suvah", "Mahah", "Janah", "Tapah" and "Satyam" are called seven Vyahruti's or Sapta Vyahrutis. These denote different worlds in Puranas. The term "Vyahruti" means pronounciation, since these above mentioned words are pronounced in Vedic rituals like Agnihotra, these are called Vyahruti's.

Fifth Anuvak states that the fourth Vyahruti "Mahah" was discovered by a Rishi called "Mahachamasya" and "Mahah" is Brahman all other Vyahrutis are its organs. This Anuvak says that "Bhooh" denotes earth, fire, Richa's and Prana. "Bhuvah" denotes space, air, Sama's and Apana. "Suvah" denotes heaven, sun, Yajus and Vyana. "Mahah" denotes sun, moon, Holy Syllable OM and food. This way, each of first four Vyahrutis become four each and in total they are 16. So all these four Vyahrutis should be meditated as all of their sixteen manifestations, this is known as famous vedic term Shodasha Kala Purusha in vedic literature. Finally this anuvak says that one who knows all sixteen manifestations of Vyahrutis knows Brahman and all gods bring gifts to him.

Sixth Anuvak tells that in the space inside the heart there exists an immortal golden being (Hiranmaya Purusha) and states that there is a subtly route through the middle of the head through which a self realized Yogi travels when he/she leaves the body and joins the all-pervading Brahman.

The Seventh Anuvak, is bit difficult to understand, It describes a meditation called "Paanktha Upasana", where Brahman is meditated upon as a set of "five" called "Paanktha". There is also a vedic meter of five lines called Paanktha. Outer world seen is classified into 3 set of five things called "Adi Bootha" and also inner body world into 3 sets of five things called "Adhyaathma". This inner-outer grouping is then meditated upon and meditator realizes both inner and outer paankta are one and the same Brahman and every thing is Paanktha. This Anuvak states that Earth (Prithvi), Space (Antariksha), Heaven (Dyuloka), Directions (North, South, etc.), Subdirections (South-East, North-West, etc..) (Worldly set of Five or Loka-Paankta) and

Fire (Agni), Air (Vayu), Sun (Aditya), Moon (Chandra), Stars (Nakshatra) (Divine-Five or Deva-Paanktha) and

Water, Herbs (Oshadhi), Trees, Sky (Akasha), Self (Atma) (Elemental Five or Dhaatu Paanktha) are called Adibhuta or Outer Elements.

Further it states that:

*Prana, Vyana, Apana, Udana, Samana (all these are Five inner Vital currents of the body) and*

*Eye, Ear, Mind, Speech (Vaak), Outer-Skin (Tvak) (Sensory-Five or Indriya-Paanktha) and*

Inner-Skin (Charma), Flesh, tendon-Ligaments, bones, Fat or Bone-Marrow (Elemental Five or Dhaatu Paanktha) are called Adhyaatma or Inner Elements.

The Rishi who realized this says, thus everything is Paankta. It further says that one who realized everything as Paanktha, completes Paanktha with Paanthks itself.

Eighth Anuvak states the greatness of Holy Syllable OM. It says that OM is Brahman and everything.

Ninth Anuvak explains the duties to be performed by persons aspiring realization of Brahman.

Tenth Anuvak is a Mantra for self-practice (Swadhyaya) or meditation. It is stated by a Rishi called Trishanku as an exclamation of awe after he realized that he is one with Brahman. It is also called "Mantramnaya" of Trishanku.

Eleventh Anuvak is a set of instructions that teacher (Acharya) gives to his desciple after completion of vedic education.

Twelveth Anuvak concludes the Shikshavalli with Santi Mantra "OM Sham no mitra.." expressing gratitude toward gods who removed obstacles for study of Upanishad as prayed for in first anuvak of this Valli.

***Brahmananda Valli:*** The statement *"Brahmavida Apnoti Param"* which means "The one who knows Brahman becomes one with Brahman" is the formula (Sutra-Vaakya) to get the high level gist of this Valli. First Anuvak starts with Shanti Mantra "OM sham no mitra" and "Shahana vavatu" pleasing gods and removing obstacles for study of Upanishad being the objective of these Mantras.

Second Anuvak starts with formula sentence "Brahmavida Apnoti Param" as stated above and also tries to define brahman succintly as "Truth, Devine Knowledge and Infinite "(*Satya Jnyanam Anantam Brahma*).

Anuvaks Second to Fifth describe that Five sheaths subtle bodies or (Five Atmans) reside in one another in human body. Starting with grosser, tangible human body called "Annamaya" or "Formed out of Food" to "Pranamaya" or "formed out of Vital life force" to "Manomaya" or "Formed out of Mind" to "Vijnyanamaya" or "One who is of Knowledge" to Final and subtle most being "Anandamaya" or one who is full of *Joy*. In Sixth and Seventh Anuvaks, some of the questions asked by a desciple are answered, such as,

"Brahman being equal to both knower and ignorant, who gets the Brahman after death, knower or ignorant and why?"

Eighth Anuvak, compares happiness of various evolved beings starting from Man to that of next higher level till Happiness of Brahman itself.

Ninth Anuvak describes that knower of Brahman doesn't repent for not having done any good because for him/her, the terms good and bad loose their meaning and he/she has equalled them with Brahman since it is the only one which is really existing.

With this completes the "Brahmanada Valli".

***Brighu Valli:*** This Valli describes how son of Varuna (The Water God) the [[*Brighu*]] obtained realization of Brahman through repeated Tapas and his father's guidance.

Rest of the part of Valli describes greatness of donating food, that is feeding the hungry. It also emphasises on greatness of Food. It says that since food is support of all life, food should not be insulted, food should not be declined.

## Chandogya Upanishad

The *Chandogya* Upanishad is one of the oldest (perhaps "the oldest"), "primary" Upanishads commented upon by Adi Shankara. It is a Mukhya Upanishad, associated with the Samaveda. It figures as number 9 in the Muktika canon of 108 Upanishads. It is part of the *Chandogya Brahmana* which has ten chapters. The first two chapters of the Brahmana deal with sacrifices and other forms of worships. The other eight constitute the Chandogya Upanishad.

Though there are more than one hundred Upanishads, ten are principal. These are known as Dashopanishads and known for their philosophical depth and because Adi Shankara wrote commentaries on them.

Along with Brhadaranyaka Upanishad the *Chandogyopanishad* is an ancient source of principal fundamentals for Vedanta philosophy. Number of references made to this Upanishad in Brahma sutras indicates special importance of this Upanishad in Vedantic philosophy. Important spiritual practices like Dahara vidya, Shandilya vidya, etc. are its speciality.

***Commentary:*** Of the available commentaries the oldest is the one written by Adi Shankara. Adi shankara has indicated that his commentary work is a brief book for those who want to know summary of this Upanishad. Commentator Ananda Giri has mentioned in his commentary that one Dramidacharya has written an elaborate and detailed commentary well before Adi Shakara. Very little is known about this Dramidacharya and his work is now lost.

Brahmanandi Tankacharya has written a brief explanation for this upanishad and Dramidacharya wrote an elaborate and detailed commentary on the work of Brahmanandi Tankacharya. Shri Ramanuja makes many references to these two scholars in his *Vedanta Sangraha* and *Shree Bhashya* both of which are commentary works.

The first and the second chapters of this Upanishad discuss the problems of liturgy and doctrine such as the genesis and significance of Aum and the meaning and names of Saman.

First chapter consists of eighteen subchapters in which the following are described:

1. Importance and greatness of universal holy syllable OM
2. Importance and greatness of Vital Life Force or Prana, and a story to explain the holiness of Prana describing

how it remained untouched by evil while all other five senses got tainted by evil.

3. Fruits of meditating on holy syllable OM

## Kena Upanishad

The *Kena* Upanishad (kenopanicad), is one of the older, "primary" Upanishads commented upon by Shankara. It is associated with the Samaveda. It figures as number 2 in the Muktika canon of 108 Upanishads.

***About the Upanishad:*** Kenopanishad derives its name from the first word *Kena*, meaning 'by whom'. It belongs to the Talavakara Brahmana of Samaveda and is therefore also referred to as Talavakara Upanishad. It has four sections, the first two in verse and the other two in prose. Adi Shankara who has written commentaries on 12 Upanishads, chose to write two commentaries on Kenopanishad. One is called Kenopanishad Pada Bhashya and the other is Kenopanishad Vakya Bhashya.

***Brief Content:*** The One Power that illumines everything and every one is indivisible. It is the Ear behind the ears, Mind behind the mind, Speech behind speech, Vital Life behind life. The ears cannot hear it; it is what makes the ears hear. The eyes cannot see it; it is what makes the eyes see. You cannot speak about it; it is what makes you speak. The mind cannot imagine it; it is what makes the mind think. It is different from what all we know; yet it is not known either. Those who feel they know Him know Him not. Those who know that anything amenable to the senses is not Brahman, they know it best. When it is known as the innermost witness of all cognitions, whether sensation, perception or thought, then it is known. One who knows thus reaches immortality.

***Story Part:*** Once the divines won a victory over the evil forces. The victory must have been credited to the power of the Absolute Brahman. Instead the divines thought it was theirs. Brahman appeared before them in a visible form of a spirit (*yaksha*) but they did not recognize the Absolute. One by one, Agni the God of fire and Vayu the God of air, came to challenge this new appearance in and tried to show off their powers. The God of Fire could not burn even the straw placed before him. The God of air could not blow even the straw placed before him. Finally Indra the God of all the divines came nearest to that spirit to find out who it is that is presenting these challenges to the divines. And before him stood Mother Goddess Herself in the name and form of Uma.

***Final Lesson:*** This is the truth of Brahman in relation to nature and Man. Whether it is the flash of lightning, or the wink of the eyes or the thinking of the mind, the power that is shown is the power of Brahman. For this reason should a man meditate upon Brahman all the time. The sudden Reality that strikes Man as the power behind everything, must be transformed into a permanent Realization.

## Mundaka Upanishad

The *Muaka* Upanishad is one of the older, "primary" (*mukhya*) Upanishads commented upòn by Shankara. It is associated with the Atharvaveda. It figures as number 5 in the Muktika canon of 108 Upanishads.

It is a Mantra-upanishad, *i.e.* it has the form of a Mantra. But, as the commentators observe, though it is written in verse, it is not, like other Mantras, to be used for sacrificial purposes. Its only object is to teach the highest knowledge, the knowledge of Brahman, which cannot be obtained either by sacrifices or by worship (Upasana), but by such teaching only as is imparted in the Upanishad.

With its beautiful style, lucid metres, serious wording, and lofty feelings each mantra of this Upanishad gives joyous reading. It might have derived its name from the word "Munda" meaning "Shaven Head".

The assumption is that if the principle thought of this Upanishad is understood, the illusions of material world will be cut like hairs in the process of shaving or because mostly monks are audience for its teachings, and since usually Hindu monks will have a shaven head, this name might have something to do with that background.

It has three chapters and each chapter is divided into sub-chapters which are called “Khanda”. In total this Upanishad has 64 Mantras.

As described in the beginning of this upanishad, it is said to be first told by Brahma to his son Atharva and Atharva taught it to Satyavaha and Satyavaha passed it to Angiras who in turn passes the knowledge to Shaunaka, dialogue between two forms content of this Upanishad.

This Upanishad divides all knowledge into two categories. The knowledge that leads to Self Realization is called Para Vidya or Divine Knowledge and everything else is called Apara Vidya or Knowledge of Material world. It is the first text to mention the six disciplines of Vedanga.

This Upanishad expounds the greatness of Para Vidya.

The Mundaka Upanishad is notable as the source of the phrase *Satyameva jayate*, the national motto of India, appearing in the Nation Emblem of four lions.

## Mandukya Upanishad

**Mandukya Upanishad** is one of the shortest Upanishads, that form the speculative metaphysical parts of the Hindu texts, the Vedas. It belongs to the Atharva Veda. It devotes itself entirely to the explanation of the mystic syllable Aum. It is in prose.

***About the Upanishad:*** For the very reason that it explains the esoteric meaning of the fundamental syllable *Aum* of Hindu spiritual tradition, the Upanishad has been extolled greatly. The Muktikopanishad, which talks about all other Upanishads, says that if a person cannot afford to study all the hundred and more Upanishads, it will be enough to read just the Mandukya Upanishad. According to Dr.S. Radhakrishnan, in this Upanishad we find the fundamental approach to the attainement of reality by the road of introversion and ascent from the sensible and changing, through the mind which dreams, through the soul which thinks, to the divine within but above the soul.

***Commentary by Gaudapada:*** The first extant commentary on this Upanishad was written by Gaudapada, before the time of Adi Shankara. This commentary, called the Mandukya-karika, is the earliest systematic exposition of the advaita point of view of Vedanta. Its importance can be gauged from the fact that when Shankara wrote his commentary on Mandukya Upanishad, as he did for ten other Upanishads, he merged the Karika of Gaudapada with the Upanishad and wrote a commentary on the Karika also.

## Detailing Aum as in Mandukya Upanishad

***Three matras:*** There are three matras in the word *aum* : 'a' as the 'u' in 'but'; 'u' as the 'u' in 'put'; and the 'm'. The term matra is used for the upper limb of Nagari characters and a syllabic instant in prosody. Esoterically, the 'a' stands for the first stage of wakefulness, where we experience in our gross body the totality of external experiences through our mind and sense organs. The 'u' stands for the dream state of sleep in which mental experiences are available, though erratically, by the mind which is the only thing which is then awake, without the help of the external sense organs or the presence of the rationalising intellect.

***Waking State and Dream State:*** The two kinds of experience, namely those of the waking state and those of the dream state, contradict each other, in the sense that a man may experience hunger in a dream though he has eaten in the waking state a few minutes earlier.

***Deep Sleep State:*** In the state of deep sleep, represented by the sound 'm', there is no consciousness of any experience; even the mind has gone to sleep. But still there is an awareness after the deep sleep is over that one has been sleeping. Mandukya Upanishad says that in the state of deep sleep, the Atman which is always present, has been the witness to the sleep of the body and it is this source from which issues the memory of sleep.

***Beyond the Three States:*** It is the Atman which is also present beyond the three states of experience. The fourth state (*turiya avastha*) (see turiya) corresponds to the silence that ensues after one has steadily pronounced aum. It is the state of no matra (amatra). In that silence Consciousness alone is present; there is nothing else. Therefore there is nothing to be cognized or be conscious of. This is the substratum of even the other three states of experience.

During the silence that follows the recitation of *aum*, one is advised to merge in that Consciousness, in fact, be that Consciousness. That Consciousness is the Atman. That is Brahman. To underscore the point that the 'fourth state' is not another 'state' of consciousness, but consciousness itself, *turiya avastha* is simply called *turiya* (the fourth).

***Gaudapada's Thesis:*** In his karika on the Upanishad, Gaudapada deals with all the outstanding problems of philosophy, such as perception, idealism, causality, truth, and reality. In turiya, he says the mind is not simply withdrawn from the objects but becomes one with Brahman, who is free from fear and who is all-round illumination. In both deep sleep and transcendental consciousness there is no consciousness of objects.

But this objective consciousness is present in an unmanifested 'seed' form in deep sleep while it is completely transcended in the *turiya*. Specifically, if one identifies the amatra state of silence with the *turiya* and meditates on it without intermission, one realizes one's self and 'there is no return for him to the sphere of empirical life', says Gaudapada.

## Prashna Upanishad

***Prashna*** Upanishad (IAST pracnopanicad) is one of the older, "primary" Upanishads commented upon by Shankara. It is a Mukhya Upanishad, associated with the Atharvaveda. It figures as number 4 in the Muktika canon of 108 Upanishads.

In Sanskrit, "Prashna" means question. This book consists of six questions and their answers, hence the name. It is in the form of question-answers, except first and last questions, all other questions are actually a group of smaller sub-questions.

As narrated in the beginning of this Upanishad, Six pupils interested in knowing divinity or Brahman come to sage Pippalada and ask questions of great spiritual importance. Pippalada asks them to take up penance of one year. Upon completion of penance, they again come to sage and ask questions, then the sage answers their questions.

The pupils who ask questions are:

1. The son of Bharadwaja the Sukesha
2. The son of Shibi the Satyakama
3. The descendant of Garga the Sauryayanee
4. The son of Ashwalayana the Kausalya
5. Bhargava of the country of Vidarbha
6. The son of Katya the Kabandhi

Each of them asks one question to Pippalada and answer(s) to it forms a chapter in the Upanishad.

***The Questions:*** The first question is asked by Kabandhi concerned with the root cause of the universe.

The second question asked by Bhargava is concerned with the supremacy of vital Force of Life or Prana over other sense organs of human being.

The third question asked by Kausalya is concerned with the origin and functioning of Vital Force of Life.

The fifth question is concerned with fruits one gets by meditating on holy syllable OM.

The sixth and last question is concerned with the being by whom all this known-unknown universe got created and who has sixteen vital things (kala) or who is called as *Shodasha Kala Purusha*.

The answers to each question are developed by Pippalada, within a perspective of great depth. The language and concepts of the answers are often abstruse and esoteric, but one may get from them glimpses that are rational and perceptive. Thus, in answering the question on the origin of life, Pippalada starts with the origin of all existence in terms of matter and energy, and develops the answer step by step, till in the last one, he gives as the most direct and immediate origin of life, the sperm of the species. The answer to the last question is, that the ultimate supreme source of all existence is Brahman. This, indeed, is the answer of all the Upanishads.

## Shvetashvatara Upanishad

The *Shvetashvatara Upanishad* is one of the older, "primary" Upanishads. It is associated with the Black Yajurveda. It figures as number 14 in the Muktika canon of 108 Upanishads. Adi Shankara has called it the "Mantra Upanishad" of the Vedic Shvetashvatara school in his commentary on Brahma sutras.

This Upanishad contains 113 mantras or verses in six chapters. In the last chapter we find the following verse.

*"Sage Shvetashvatara got this knowledge of Brahman which is very sacred and revered by many great sages by his penance and god's grace and taught it very well to his disciples"*

By this verse we can make out that this Upanishad was due to a sage called "Shvetashvatara" or his line of ancient spiritual teachers. The name "Shvetashvatara" is not uncommon in vedic literature. It means "White Mule". Mule was a prized animal in ancient vedic India. The person who owns white horse is called "Shvetashva" and the one who owns white mule can be called "Shvetashvatara". We can notice that Arjuna is called as "Shvetashva" in epic Mahabharata. In Rigveda we find a sage's name as "Shyavashva" meaning "One who owns a black horse". This Upanishad has acquired a high place among the Hindu Kruti.

***Commentary Works:*** There is one commentary available on this upanishad that is said to be written by Adi Shankara. But if we compare it with other commentary works of same author, there is some doubt if it is indeed composed by Adi Shankara. There are other three commentators namely Vijnanatma, Shankarananda and Narayana Thirtha.

***Time and Origin:*** Deciding time of the origin of this Upanishad in not easy because, apart from very ancient content, there are also other verses which look like they were added in at a later point of time. Many Mantras in this upanishad can also be seen in Rigveda, Black Yajurveda, White Yajurveda. Though it may not be as old as liturgic Upanishads like Chandogya or Brhadaranyaka Upanishad, we can say it belongs to time line of poetic group of Upanishads like Katha Upanishad, Eshavasya or Mandukya Upanishad.

***Special Features:*** This Upanishad is just as much tough to understand as much it is wonderful. It has following specialities which arouse curiosity in the reader of vedic literature.

1. The Supreme God or Brahman is called by various names such as Shiva, Rudra, etc. This feature may give rise to an assumption that it may be a Shaiva Upanishad. But such an assumption is not right because, at the time of this Upanishad, Shaiva Agamas weren't there. Also, at that time the supreme God head or Brahman usedbe called by different names each indicating a particular manifestation of Brahman. Just as names Shiva or Rudra are used to refer Brahman, names such as Vayu, Aditya or Agni are also used for same purpose in this Upanishad. More over, if this Upanishad is indeed a Shaiva Upanishad, other sects of Hinduism such as Vaishnavas wouldn't have quoted its verses/ mantras as authority in their respective treatises.
2. Second speciality is concept of "*Devotion*" or Bhakti. In other Upanishads, the concept of devotion (or Bhakti) is indirectly voiced in the form of "*Upasana*". But here, without leaving any room for guesswork, it is explicitly mentioned. The word "*Para Bhakti*" is explicitly used at the end of sixth chapter. There are many words which voice submission to God or Brahman (words such as "*Sharanam*", "*Prapadye*"). This concept of devotion at later times found profound and wide meanings in *Bhakti Sutras* and other treatises on Bhakti or devotion.
3. Third speciality of this upanishad lies in its giving importance to the *form* fullness (or *Moortitva*) of God or Brahman which is usually described as *formless* in other important upanishads. Since it is difficult to concentrate mind on and/or show devotion on *formless* Brahman, it is only natural that this book goes ahead and tries to ascribe various *forms* to God. While expounding on devotion, it also ascribes various characteristics or manifestations to God, such as,

mentioning that God has knowledge and power. This upanishad also presents that God or Brahman as creator and sustainer of the universe, and while trying to describe various powers of God, to drive home the point, makes use of similes such as God having thousands of heads-to denote endless knowledge of God, thousands of eyes-to denote God as universal witness for everything that is going on in the universe and God having thousands or simply innumerable number of feet—to indicate Gods omnipresence. This Upanishad mentions that, God or "*Parama Purusha*" is shining in his (or "its" since upanishads do not ascibe gender to Brahman) glory beyond the darkness of ignorance or "*Tamas* ". Gods characteristic is "*Maya*" or illusion of the world. But God is not bound by his Maya as humans are, because he has conquered it and became "*Maheshvara*", and he (or it) is capable of giving salvation to human beings.

4. Fourth speciality is use of words such as "*Samkhya*", "*Kapila*", "*yoga*", "*Prakriti*". Some scholars debate that this indicates assimilation of Saankhya-Yoga-Darshana school of thought into Vedanta happened at some point of time. If this view is right, it means that Vedantic Upanishads were originated later than Saankhya-Yoga-Darshana school of thought. This view is not right because of following reasons.
   (i) Just because certain definitive words are similar, it doesn't imply Upanishads are of later origin than Darshana schools. The definitive words like "*Samkhya*, *prakriti,* etc. are always used in Vedas earlier than Darshana schools. So, these specific words are taken from Vedanta and later used by proponents of Darshana schools.
   (ii) The proponents of Darshanas like sage kapila describe in their treatises that they are interpreting the Vedanta, so it means Vedantic Upanishads are of earlier origin than these schools of thought.
   (iii) In this Upanishad, we don't see any assimilation

of principles of those schools but to contrary, the thoughts that are rejected by those schools are expounded and accepted using the same words and definitions of those schools.

The philosophical jargon of Samkhya school of thought originated in vedas well before birth of Saamkhya. The word saamkhya originates from word "samkhya" which means counting or understanding. There are many verses in this upanishad that have some counting of things in them that is why word "samkhya" got another meaning as knowledge. While Yoga is practical Sammkhya is intellectual. While samkhya is purely dualistic. Upanishads are clearly non-dualistic.

In Samkhya the "Purusha" or self is different from "Prakriti" or Nature but in Upanishadic thought, "Prakriti"(Nature) is God derived, or God dependent force, specifically this Upanishad refers to it as "*Devaatma Shakti*", which can be roughly translated as "Gods inner force that manifests itself as nature (*Prakriti*)". One hymn of this Upanishad says,

*"Sages engaged in deep meditation realized secret 'Atma Shakti' of God."*

It says this *Atma Shakti* of God is hidden secretly in the meditators' own inner nature.

Regarding the usage of the word "Kapila" is concerned, if we see the contexts of hyms that have come before and later of the hymn which uses word kapila, it is clear that it does not denote sage kapila who is proponent of Samkhya. The hymn 5.2 which uses word "kapila" should be seen in the context with other hyms. 4th hymn of 3rd chapter says,

*"All knower Rudra gave birth to Hiranyagarbha"*

12th hymn of 4th chapter says

*"Rudra saw the birth of Hiranyagarbha"*

later, 18th hymn of 6th chapter says,

*"Parameshvara created Brahma"*

Now when seen in context of these hymns, the verse 5.2, which uses the word kapila says,

*"God saw birth of All-knowing Kapila"*

In Sanskrit, word *kapila* means "golden-coloured". *Hiranya garbha* means "one embodying golden core". In Upanishads *Hiranya garbha* is also called by other names such as Prajapati, Brahma, etc. based on this context relation, we can conclude this Upanishad when it uses word kapila it means *Hiranya garbha* rather than Samkhya school proponent.

***Poetic Style:*** Normally, Upanishads are sources of serious philosophical thought. But this Upanishad differs with other Upanishads in being same principles explained in simple, easy-going and poetic way, whereever sage svhetashvatara has his independent hymns, he sees them in a beautiful, heart-catching poetic way, he is not only a seer of Mantras, but also a poet from heart. Here are few examples.

While trying to describe omnipresence of Brahman, Hymn 4.2 says,

> *"You are woman, you are man, you are boy and you are girl, you are the shivering old man helped by stick, you are born in the form of this world"*

Hymn 4.4 says

> *"You are blue-coloured butterfly, green eyed parrot, lightning cloud, you are the seasons, and seas, you are the one without any beginning, you are omnipresent, all the worlds are born out of you"*

## The Anukramanis

The Anukramanis or Indices of the Rigveda provide us with the most basic information about each of the 1028 hymns of the Rigveda:

a. The Rishi or composer of each hymn or verse.
b. The Devata or deity of each hymn or verse.
c. The Chhanda or metre of each hymn or verse.

For the purpose of our historical analysis of the Rigveda, we will be concerned only with the index which deals with the most undeniably historical aspect of the Rigveda: the index of Rishis which provides us with details about the living and breathing historical personalities who composed the hymns.

The Rigveda consists of 10 Mandala or Books, which contain 1028 suktas or hymns, consisting of 10552 mantras or verses as follows:

| *Mandala N* | *No. of Hymns* | *No. of verses* |
|---|---|---|
| I | 191 | 2006 |
| II | 43 | 429 |
| III | 62 | 617 |
| IV | 58 | 589 |
| V | 87 | 727 |
| VI | 75 | 765 |
| VII | 104 | 841 |
| VIII | 103 | 1716 |
| IX | 114 | 1108 |
| X | 119 | 1754 |
| Total | 1028 | 10552 |

The Anukramanis give us details, regarding these hymns, which are *so basic and indispensable* that it is *inconceivable* that any serious scholar could consider it possible to analyse the hymns *without* taking the Anukramanis as the very basis for his analysis.

But, ironically, not only are the Anukramanis generally ignored by the scholars, but this ignorance of, and indifference to, the details contained in the Anukramanis is even flaunted by them.

Consider the following statements by eminent scholars who consider themselves qualified to make pronouncements on Rigvedic history:

B.K. Ghosh: "The first Mandala falls naturally into two parts: the first fifty hymns have the Kanvas as authors like the eighth Mandala...".

Actual fact: I.1-11, 24-30 (eighteen hymns) are by Vishvamitra.

I.31-35 (five hymns) are by Angirases

I.12-23, 36-50 (twenty-seven hymns) are by Kanvas

DD Kosambi: "The principal Vedic god is Agni, the god of fire; more hymns are dedicated to him than to any other. Next in importance comes Indra."

Actual fact: The ratio between the number of hymns and verses to the two gods, by any count, is Indra: Agni = 3:2.

The flippant attitude of these scholars towards factual details, when it comes to Rigvedic studies, is underlined by the nature of Kosambi's error: he misinterprets the fact that hymns to Agni are generally *placed before* hymns to Indra, to mean that there are more hymns to Agni than to Indra!

Maurice Bloomfield, in his invaluable work on Rigvedic Repetitions (*i.e.* verses, verse-sections or phrases, which occur more than once in the Rigveda) claims that these repetitions prove the falsity or dubiousness of the information contained in the Anukramanis:

Under the title "Untrustworthiness of Anukramani-statements Shown by the Repetitions", Bloomfield remarks that "the statements of the Sarvanukramani.... betray the dubiousness of their authority in no particular more than in relation to the repetitions.... the Anukramani finds it in its heart to assign, with unruffled insouciance, one and the same verse to two or more authors, or to ascribe it to two or more divinities, according as it occurs in one book or another, in one connexion or another. The AprI stanzas 3.4.8-11 = 7.2.8-11 are ascribed in the third book to Vishvamitra Gathina, in the seventh book to Vasishtha Maitravaruni."

However:

1. The repetitions do *not* disprove the authenticity of the Anukramanis:
    a. The repetitions in the Rigveda are representative of a regular phenomenon in Classical and liturgical literature throughout the world. Consider for example what Gilbert Murray says about similar repetitions in Greek literature: "descriptive phrases...... are caught up ready made from a store of such things: perpetual epithets, front halves of lines, back halves of lines, whole lines, if need be,

and long formulae. The stores of the poets were full and brimming. A bard need only put in his hand and choose out a well-sounding phrase. Even the similes are ready-made." Quoting this, B.K. Ghosh notes: "All this may be maintained, *mutatis mutandis*, also of Rigvedic poetry."

In the case of the Rigveda it is significant that *every single* repetition pertains to a literary or liturgical phrase. In fact, the more literary or liturgical the reference, the more the likelihood of repetitions: the longest repetition of three consecutive verses is in the liturgical AprI-suktas of the Vishvamitras and Vasishtha: III.4.8-11 = VII.2.8-11.

*Not a single* repetition pertains to any historical reference: even when the same historical reference is found in four different verses, the phrasing is different: I.53.10; II.14.7; VI.18.13; VIII. 53.2.

Therefore, regardless of the number of verses or verse-sections common to any two hymns ascribed to two different Rishis, the hymns in question have to be regarded as compositions of the two Rishis to whom they are ascribed: that one Rishi has borrowed from the composition of the other is no criterion in judging the correctness of the Anukramanis.

b. The AprI-suktas of the Vishvamitra and Vasishtha contain the longest repetitions, of three verses, in common: III.4.8-11 = VII. 2.8-11. Bloomfield points to these particular repetitions as evidence in support of his contention that the repetitions disprove the correctness of the Anukramanis. But, ironically, it is these very repetitions which disprove the correctness of his *contention*.

The composers of the Rigveda were members of ten priestly families, and each family had its own AprI-sukta composed by a member of the family. In later times, during the performance of any sacrifice, at the point where an AprI-sukta was to be recited, the conducting Rishi was required to recite the AprI-sukta of his own family.

The AprI-sukta of the Vishvamitra was therefore undoubtedly composed by a Vishvamitra, and that of the Vasishtha by a Vasishtha. If these two hymns contain repetitions in common, it constitutes the ultimate proof that repetitions in common are no evidence of two hymns *not* having been composed by two different Rishis.

2. There is no logical reason to doubt the authenticity of the authorship ascriptions in the Anukramanis, which are corroborated by:
   a. The very existence of the Anukramanis as a part and parcel of the Rigvedic text from the most ancient times.
   b. The very division of the Rigveda into Mandala, many of which are family Mandala.
   c. The uniformity of style in hymns ascribed to single Rishis or families (eg. Parucchepa).
   d. The common refrains occurring in the concluding verses of hymns ascribed to certain Rishis or families (eg. Kutsa).
   e. The common contexts in hymns ascribed to certain Rishis or families (eg. the repeated references to Sudas in hymns by Vasishtha).
   f. Specific statements within the hymns, where the composers refer to themselves by name.
   g. Most important of all, the perfectly logical way in which an analysis of the historical references in the hymns, as we shall demonstrate in this book, produces a pattern of historical correspondences and inter-relationships which fits in perfectly with the ascriptions in the Anukramanis.

With this, we may now turn to the actual details given in the Anukramanis regarding the names of the composers of the different hymns in the Rigveda:

| | |
|---|---|
| | ***Mandala I (191 hymns)*** |
| 1-10 | Madhucchandas Vishvamitra |
| 11 | Jeta Madhucchandas |
| 12-23 | Medhatithi Kanva |
| 24-30 | Sunahsepa Ajigarti later Devarata Vishvamitra |
| 31-35 | Hiranyastupa Angiras |
| 36-43 | Kanva Ghaura |
| 44-50 | Praskanva Kanva |
| 51-57 | Savya Angiras |
| 58-64 | Nodhas Gautama |
| 65-73 | Parasara Saktya |
| 74-93 | Gautama Rahugana |
| 94-98 | Kutsa Angiras |
| 99 | Kasyapa Marica |
| 100 | Rjrasva Varsagira |
| 101-115 | Kutsa Angiras |
| 116-126 | Kaksivan Dairghatamas |
| 127-139 | Parucchepa Daivodasi |
| 140-164 | Dirghatamas Aucathya |
| 165-191 | Agastya Maitravaruni |

| | |
|---|---|
| | ***Mandala II (43 hymns)*** |
| 1-3 | Grtsamada Saunahotra, *later* Grtsamada Saunaka |
| 4-7 | Somahuti Bhargava |
| 8-26 | Grtsamada Saunahotra, *later* Grtsamada Saunaka |
| 27-29 | Kurma Gartsamada |
| 30-43 | Grtsamada Saunahotra, *later* Gartsamada Saunaka |

| | |
|---|---|
| | ***Mandala III (62 hymns)*** |
| 1-12 | Vishvamitra Gathina |
| 13-14 | Rsabha Vishvamitra |
| 15-16 | UtkIla Katya |
| 17-18 | Kata Vishvamitra |
| 19-22 | Gathin Kausika. |
| 23-35 | Vishvamitra Gathina |
| 36 | Vishvamitra Gathina, Ghora Angiras |
| 37 | Vishvamitra Gathina |
| 38 | Vishvamitra Gathina, Prajapati Vishvamitra/Vacya |
| 39-53 | Vishvamitra Gathina |
| 54-56 | Prajapati Vishvamitra/Vacya |
| 57-61 | Vishvamitra Gathina |
| 62 | Vishvamitra Gathina, Jamadagni Bhargava |
| | ***Mandala IV (58 hymns)*** |
| 1-42 | Vamadeva Gautama |
| 43-44 | PurumIlha Sauhotra, AjamIlha Sauhotra |
| 45-58 | Vamadeva Gautama |
| | ***Mandala V (87 hymns)*** |
| 1 | Budha/Gavisthira Atreya |
| 2 | Kumara/Vrsa Jana Atreya |
| 3-6 | Vasusruta Atreya |
| 7-8 | Isa Atreya |
| 9-10 | Gaya Atreya |
| 11-14 | Sutambhara Atreya |
| 15 | Dharuna Angiras |
| 16-17 | Puru Atreya |
| 18 | Dvita Atreya |

| | |
|---|---|
| 19 | Vavri Atreya |
| 20 | Prayasvanta Atreya |
| 21 | Sasa Atreya |
| 22 | Visvasaman Atreya |
| 23 | Dyumna Visvacarsani Atreya |
| 24 | Bandhu, Subandhu, Srutabandhu, Viprabandhu (Gaupayanas) |
| 25-26 | Vasuyava Atreya |
| 27 | Atri Bhauma |
| 28 | Visvavara Atreyi |
| 29 | GaurivIti Saktya |
| 30 | Babhru Atreya |
| 31 | Avasyu Atreya |
| 32 | Gatu Atreya |
| 33-34 | Samvarana Prajapatya |
| 35-36 | Prabhuvasu Angiras |
| 37-43 | Atri Bhauma |
| 44 | Avatsara Kasyapa, various Atreyas |
| 45 | Sadaprna Atreya |
| 46 | Pratiksatra Atreya |
| 47 | Pratiratha Atreya |
| 48 | Pratibhanu Atreya |
| 49 | Pratiprabha Atreya |
| 50-51 | Svasti Atreya |
| 52-61 | Syavasva Atreya |
| 62 | Srutavida Atreya |
| 63-64 | Arcananas Atreya |
| 65-66 | Ratahavya Atreya |
| 67-68 | Yajata Atreya |
| 69-70 | Urucakri Atreya |

| | |
|---|---|
| 71-72 | Bahuvrkta Atreya |
| 73-74 | Paura Atreya |
| 75 | Avasyu Atreya |
| 76-77 | Atri Bhauma |
| 78 | Saptavadhri Atreya |
| 79-80 | Satyasravas Atreya |
| 81-82 | Syavasva Atreya |
| 83-86 | Atri Bhauma |
| 87 | Evayamarut Atreya |

***Mandala VI (75 hymns)***

| | |
|---|---|
| 1-30 | Bharadvja Barhaspatya |
| 31-32 | Suhotra Bharadvaja |
| 33-34 | Sunahotra Bharadvaja |
| 35-36 | Ǹara Bharadvaja |
| 37-43 | Bharadvaja Barhaspatya |
| 44-46 | Samyu Barhaspatya |
| 47 | Garga Bharadvaja |
| 48 | Samyu Barhaspatya |
| 49-52 | Rjiśvan Bharadvaja |
| 53-74 | Bharadvaja Barhaspatya |
| 75 | Payu Bharadvaja |

***Mandala VII (104 hymns)***

| | |
|---|---|
| 1-31 | Vasishtha Maitravaruni |
| 32 | Vasishtha Maitravaruni Shakti Vasishtha |
| 33-100 | Vasishtha Maitravaruni |
| 101-102 | Vasishtha Maitravaruni, Kumara Agneya |
| 103-104 | Vasishtha Maitravaruni |

***Mandala VIII (103 hymns)***

| | |
|---|---|
| 1 | Pragatha Kanva, Medhatithi Kanva, Medhyatithi Kanva |
| 2 | Medhatithi Kanva, Priyamedha Angiras |
| 3 | Medhyatithi Kanva |
| 4 | Devatithi Kanva |
| 5 | Brahmatithi Kanva |
| 6 | Vatsa Kanva |
| 7 | Punarvatsa Kanva |
| 8 | Sadhvamsa Kanva |
| 9 | SaSakarNa Kanva |
| 10 | Pragatha Kanva |
| 11 | Vatsa Kanva |
| 12 | Parvata Kanva |
| 13 | Narada Kanva |
| 14-15 | Gosuktin Kanva, Asvasuktin Kanva |
| 16-18 | Irimbitha Kanva |
| 19-22 | Sobhari Kanva |
| 23-25 | Visvamanas Vaiyasva |
| 26 | Visvamanas Vaiyasva, Vyasva Angiras |
| 27-31 | Manu Vaivasvata or Kasyapa MarIca |
| 32 | Medhatithi Kanva |
| 33 | Medhyatithi Kanva |
| 34 | Nipatithi Kanva |
| 35-38 | Syavasva Atreya |
| 39-41 | Nabhaka Kanva |
| 42 | Nabhaka Kanva, Arcananas Atreya |
| 43-44 | Virupa Angiras |
| 45 | Trisoka Kanva |
| 46 | Vasa Asvya |

| | |
|---|---|
| 47 | Trita Aptya |
| 48 | Pragatha Kanva |
| 49 | Praskanva Kanva |
| 50 | Pustigu Kanva |
| 51 | Srustigu Kanva |
| 52 | Ayu Kanva |
| 53 | Medhya Kanva |
| 54 | Matarisvan Kanva |
| 55 | Krsa Kanva |
| 56 | Prsadhra Kanva |
| 57-58 | Medhya Kanva |
| 59 | Suparna Kanva |
| 60-61 | Bharga Pragatha |
| 62-65 | Pragatha Kanva |
| 66 | Kali Pragatha |
| 67 | Matsya Sammada or Manya Maitravaruni |
| 68-69 | Riyamedha Angiras |
| 70 | Puruhanman Angiras |
| 71 | Suditi PurumIlha |
| 72 | Haryata Pragatha |
| 73-74 | Gopavana Atreya |
| 75 | Virupa Angiras |
| 76-78 | Kurusuti Kanva |
| 79 | Krtnu Bhargava |
| 80 | Ekadyu Naudhasa |
| 81-83 | Usidin Kanva |
| 84 | Usana Kavya, |
| 85 | Krsna Angiras |
| 86 | Krsna Angiras, Visvaka Karsni |

| | |
|---|---|
| 87 | Krsna Angiras, DyumnIka Vasishtha, Priyamedha Angiras |
| 88 | Nodhas Gautama |
| 89-90 | Nrmedha Angiras, Purumedha Angiras |
| 91 | Apala Atreyi |
| 92-93 | Sukaksa Angiras |
| 94 | Vindu Angiras, Putadaksa Angiras |
| 95-96 | Tirasci Angiras |
| 97 | Rebha Kasyapa |
| 98-99 | Nrmedha Angiras |
| 100 | Nema Bhargava |
| 101 | Jamadagni Bhargava |
| 102 | Prayoga Bhargava, Agni Barhaspatya |
| 103 | Sobhari Kanva |

***Mandala IX (114 hymns)***

| | |
|---|---|
| 1 | Madhucchandas Vishvamitra |
| 2 | Medhatithi Kanva |
| 3 | Sunahsepa Ajigarti |
| 4 | Hiranyastupa Angiras |
| 5-24 | Asita Kasyapa, Devala Kasyapa |
| 25 | Drlhacyuta Agastya |
| 26 | Idhmavaha Darlhacyuta |
| 27 | Nrmedha Angiras |
| 28 | Priyamedha Angiras |
| 29 | Nrmedha Angiras |
| 30 | Bindu Angiras |
| 31 | Gotama Rahugana |
| 32 | Syavasva Atreya |
| 33-34 | Trita Aptya |
| 35-36 | Prabhuvasu Angiras |

| | |
|---|---|
| 37-38 | Rahugana Angiras |
| 39-40 | Brhanmati Angiras |
| 41-43 | Medhatithi Kanva |
| 44-46 | Ayasya Angiras |
| 47-49 | Kavi Bhargava |
| 50-52 | Ucathya Angiras |
| 53-60 | Avatsara Kasyapa |
| 61 | Amahiyu Angiras |
| 62 | Jamadagni Bhargava |
| 63 | Nidhruvi Kasyapa |
| 64 | Kasyapa Marica |
| 65 | Jamadagni Bhargava |
| 66 | Sata Vaikhanasa |
| 67 | Saptarsis, Pavitra Angiras |
| 68 | Vatsapri Bhalandana |
| 69 | Hiranyastupa Angiras |
| 70 | Renu Vishvamitra |
| 71 | Rsabha Vishvamitra |
| 72 | Harimanta Angiras |
| 73 | Pavitra Angiras |
| 74 | Kaksivan Dairghatamas |
| 75-79 | Avi Bhargava |
| 80-82 | Asu Bharadvaja |
| 83 | Pavitra Angiras |
| 84 | Prajapati Vacya |
| 85 | Vena Bhargava |
| 86 | Atri Bhauma, Gartsamada Saunaka, Akrsta Masa, Sikata Nivavari, Prsni Aja |
| 87-89 | Usana Kavya |

| | |
|---|---|
| 90 | Vasishtha Maitravaruni |
| 91-92 | Kasyapa Marica |
| 93 | Nodhas Gautama |
| 94 | Kanva Ghaura |
| 95 | Praskanva Kanva |
| 96 | Pratardana Daivodasi |
| 97 | Vasishtha Maitravaruni, Indrapramati Vasishtha, Vrsagana Vasishtha, Manyu Vasishtha, Upamanyu Vasishtha, Vyaghrapada Vasishtha, Shakti Vasishtha, Karnasrut Vasishtha, Mrlika Vasishtha, Vsukra Vasishtha, Parasara Saktya, Kutsa Angiras. Ambarisa Varsagira, Rjisvan Angiras |
| 98 | Rebhasunu Kasyapas |
| 99-100 | Andhigu Syavasvi, Yayati Nahusa, Nahusa |
| 101 | Manava, Manu Samvarana, Prajapati Vishvamitra. Trita Aptya Dvita Aptya |
| 102 | Parvata Kanva, Narada Kanva |
| 103 | Agni Caksusa, Caksu Manava, Manu Apsava |
| 104-105 | Saptarsis |
| 106 | Gauriviti Saktya, Shakti Vasishtha, Uru Angiras, |
| 107 | Rjisvan Bharadvaja |
| 108 | Agni Dhisnya Aisvaraya Tryaruna Traivrsna, Trasadasyu Paurukutsa |
| 109 | Ananata Parucchepi |
| 110 | Sisu Angiras |
| 111 | Kasyapa Marica |

***Mandala X (191 hymns)***

| | |
|---|---|
| 1-7 | Trita Aptya |
| 8 | Trisiras Tvastra |
| 9 | Trisiras Tvastra, Sindhudvipa Ambarisa |
| 10 | Yama Vaivasvata, Yami Vaivasvati |
| 11-12 | Havirdhana Angi |
| 13 | Vivasvan Aditya |
| 14 | Yama Vaivasvata |
| 15 | Sankha Yamayana |
| 16 | Damana Yamayana |
| 17 | Devasravas Yamayana |
| 18 | Sankusuka Yamayana |
| 19 | Matitha Yamayana, or Bhrgu, or Cyavana Bhargava |
| 20-26 | Vimada Aindra, Vasukrt Vasukra |
| 27-29 | Vasukra Aindra |
| 30-34 | Kavasa Ailusa |
| 35-36 | Lusa Dhanaka |
| 37 | Abhitapa Saurya |
| 38 | Indra Muskavan |
| 39-40 | Ghosa Kaksivati |
| 41 | Suhastya Ghauseya |
| 42-44 | Krsna Angiras |
| 45-46 | Vatsapri Bhalandana |
| 47 | Saptagu Angiras |
| 48-50 | Indra Vaikuntha |
| 51-53 | Agni Saucika |
| 54-56 | Brhaduktha Vamadeva |
| 57-60 | Bandhu, Subandhu, Srutabandhu, Viprabandhu (Gaupayanas) |

| | |
|---|---|
| 61-62 | Nabhanedistha Manava |
| 63-64 | Gaya Plata |
| 65-66 | Vasukarna Vasukra |
| 67-68 | Ayasya Angiras |
| 69-70 | Sumitra Vadhryasva |
| 71-72 | Brhaspati Angiras |
| 73-74 | Gauriviti Saktya |
| 75 | Sindhuksit Praiyamedha |
| 76 | Jaratkarna Sarpa Airavata |
| 77-78 | Syumarasmi Bhargava |
| 79-80 | Agni Saucika or Sapti Vajambhara |
| 81-82 | Vishvamitra Bhauvana |
| 83-84 | Manyu Tapasa |
| 85 | Surya Savitri |
| 86 | Vrsakapi Aindra, Indra, Indrani |
| 87 | Payu Bharadvaja |
| 88 | Murdhanvan Vamadeva |
| 89 | Renu Vishvamitra |
| 90 | Narayana |
| 91 | Aruna Vaitahavya |
| 92 | Saryata Manava |
| 93 | Tanva Parthya |
| 94 | Arbuda Kadraveya Sarpa |
| 95 | Pururavas Aila, Urvasi |
| 96 | Baru Angiras, Sarvahari Aindra |
| 97 | Bhisag Atharvana |
| 98 | Devapi Arstisena |
| 99 | Vamra Vaikhanasa |
| 100 | Duvasyu Vandana |
| 101 | Budha Saumya |

| | |
|---|---|
| 102 | Mudgala Bharmyasva |
| 103 | Apratiratha Aindra |
| 104 | Astaka Vishvamitra |
| 105 | Sumitra Kautsa, Durmitra Kautsa |
| 106 | Bhutamsa Kasyapa |
| 107 | Divya Angiras, Daksina Prajapatya |
| 108 | Sarama, Panis |
| 109 | Juhu Brahmajaya |
| 110 | Rama Jamadagnya, Jamadagni Bhargava |
| 111 | Astadamstra Vairupa |
| 112 | Nabhahprabhedana Vairupa |
| 113 | Sataprabhedana Vairupa |
| 114 | Sadhri Vairupa |
| 115 | Upastuta Varstihavya |
| 116 | Agniyuta Sthaura |
| 117 | Bhiksu Angiras |
| 118 | Uruksaya Angiras |
| 119 | Laba Aindra |
| 120 | Brhaddiva Atharvana |
| 121 | Hiranyagarbha Prajapatya |
| 122 | Citramaha Vasishtha |
| 123 | Vena Bhargava |
| 124 | Agni, Varuna, Soma |
| 125 | Vak Ambhrni |
| 126 | Amhomuk Vamadeva |
| 127 | Kusika Saubhara, Ratri Bharadvaji |
| 128 | Vihavya Angiras |
| 129 | Prajapati Paramesthin |
| 130 | Yajna Prajapatya |

| | |
|---|---|
| 131 | Sukirti Kaksivata |
| 132 | Sakaputa Narmedha |
| 133 | Sudas Paijavana |
| 134 | Mandhata Yauvanasva |
| 135 | Kumara Yamayana |
| 136 | Juti, Vatajuti, Viprajuti, Vrsanaka, Karikrata, Etasa, Rsyasrnga (Vatarasanas) |
| 137 | Saptarsis |
| 138 | Anga Aurava |
| 139 | Visvavasu Devagandharva |
| 140 | Agni, Pavaka |
| 141 | Agni Trpasa |
| 142 | Sarnga, Jaritr, Drona, Sarisrkva, Stambhamitra Atri Sankhya |
| 143 | Urdhvasadman Yamayana |
| 144 | Indrani |
| 145 | Devamuni Airammada |
| 146 | Suvedas Sairisi |
| 147 | Prthu Vainya |
| 148 | Arcan Hairanyastupa |
| 149 | Mrlika Vasishtha |
| 150 | Sraddha Kamayani |
| 151 | Sasa Bharadvaja |
| 152 | Indramatara Devajamaya |
| 153 | Yami Vaivasvati |
| 154 | Sirimbitha Bharadvaja |
| 155 | Ketu Agneya |
| 156 | Bhuvana Aptya, Sadhana Aptya |
| 157 | Caksu Saurya |
| 158 | Saci Paulomi |

159 Purana Vishvamitra
160 Yaksmanasana Prajapatya
161 Raksoha Brahma
162 Vivrha Kasyapa
163 Pracetas Angiras
164 Kapota Nairrta
165 Rsabha Vairaja Sakvara
166 Vishvamitra, Jamadagni
167 Anila Vatayana
168 Sabara Kaksivata
169 Vibhrat Saurya
170 Ita Bhargava
171 Samvarta Angiras
172 Dhruva Angiras
173 Abhivarta Angiras
174 Urdhvagrava Arbuda
175 Sunu Arbhava
176 Patanga Prajapatya
177 Aristanemi Tarksya
178 Sibi Ausinara, Pratardana Kasiraja, Vasumanas
179 Rauhidasva Jaya Aindra
180 Pratha Vasishtha, Sapratha Bharadvaja,
181 Gharma Saurya Tapurmurdhan Barhaspatya
182 Prajavan Prajapatya
183 Visnu Prajapatya
184 Satyadhrti Varuni
185 Ula Vatayana
186 Vatsa Agneya

| | |
|---|---|
| 187 | Syena Agneya |
| 188 | Sarparajni |
| 189 | Aghamarsana Madhucchandas |
| 190 | Samvanana Angiras |

There are obviously corruptions in the Anukramanis in the form of ascriptions to fictitious composers. This is particularly the case in Mandala X, where a large number of hymns are ascribed to composers whose names, or patronyms/epithets, or both, are fictitious.

However, in the first eight Mandala, except in the case of one single hymn (VIII.47), it is very easy to identify the actual composer (by which we mean the Rishi who actually composed the hymn, or his eponymous ancestor to whose name the hymn is to be credited as per the system followed in the particular Mandala) of a hymn ascribed to a fictitious composer.

Hence, in our listing of the composers of the first eight Mandalas, we have replaced the fictitious names in the Anukramanis with the names of the actual composers, whose identity is clear from those same Anukramanis. In all these cases, the actual composer is the Rishi of the hymn or the Rishi of the Mandala. The hymns in question are:

(1) Hymns where the entire hymn, or verses therein, are ascribed solely (in III.23 and IV.42) or alternatively (in the others) to Rishis or kings who are referred to within the hymns by the actual composer:

| ***Hymn*** | ***Fictitious Composers*** | ***Actual Composer*** |
|---|---|---|
| I.100 | Ambarisa, Sahadeva, Bhayamana, Suradhas | Rjrasva |
| I.105 | Trita Aptya | Kutsa |
| I.126 | Bhavayavya, Romasa | Kaksivan |
| III.23 | Devasravas, Devavata | Vishvamitra |
| IV.42 | Trasadasyu Paurukutsa | Vamadeva |
| V.27 | Trasadasyu, Tryaruna, Asvamedha | Atri |
| VI.15 | Vitahavya | Bharadvaja |
| VIII.1 | Asanga, Sasvati | Medhatithi |
| VIII.34 | Vasurocis | Nipatithi |

(2) Dialogue hymns, in some of which verses are ascribed to Gods and even rivers:

| *Hymn* | *Fictitious Composers* | *Actual Composer* |
|---|---|---|
| I.165 | Indra, Maruts, (epon.) Agastya | Agastya |
| I.170 | Indra, (epon.) Agastya | Agastya |
| I.179 | (epon.) Agastya, Lopamudra, a pupil | Agastya |
| III.33 | (epon.) Vishvamitra, Rivers | Vishvamitra |
| IV.18 | (epon.) Vamadeva, Indra, Aditi | Vishvamitra |

(3) Hymns which are ascribed alternatively to the actual composers and to their remote ancestors:

| *Hymn* | *Fictitious Composers* | *Actual Composer* |
|---|---|---|
| III.31 | Kusika Aisirathi | Vishvamitra Gathina |
| VIII.27-31 | Manu Vaivasvata | Kasyapa Marica |
| VIII.71 | Purumilha Angiras | Suditi Purumilha |

# Bibliography


Aiyar, R.: *Sri Sanakra's Gita Bhashya*, Ramakrishna Math, Madras, 1983.

Ananda, K.: *The Dance of Shiva*, Oxford, New York, 1957.

Apte, V. M.: *Social and Religious Life in the Grihya Sutras*, Rawat Press, Bombay, 1954.

Austin, B.: *Guide to Indian Philosophy*, G. K. Hall, Boston, 1988.

Ballard, G.: *Philosophy at the Crossroads*, Louisiana State University Press, Baton Rouge, 1971.

Basham, A.L.: *The Origin and Development of Classical Hinduism*, Oxford University Press, New York, 1989.

Burgess, J.: *The Chronology of Modern India*, MacMillan, London, 1984.

Chand, K.: *Lord Mahavira and His Times,* Willard R. Trask. Princeton, 1969.

Charles, S.: *Technology and History*, Elek Science, London, 1974.

Chunder, S.: *The Book of Pilgrimages*, Sai Kalyani Press, Madras, 1959.

Cromwell, S.: *Ram Mohan Roy: His Era and Ethics*, Acharya Press, New Delhi, 1984.

Das, V.: *Hindu Rashtra*, Vishva Hindu Parishad, Delhi, 1990.

David, J.: *Buddhist Philosophy: A Historical Analysis*, University of Hawaii Press, Honolulu 1976.

Deutsch, E.: *Advaita Vedanta: a Philosophical Reconstruction*, East-West Center Press, Honolulu, 1969

Diana, L.: *Darsan: Seeing the Divine Image in India,* Advaita Ashrama, Calcutta, 1966.

Eliot, D.: *Advaita Vedanta: A Philosophical Reconstruction*, The University Press of Hawaii, Honolulu, 1962.

Fergusson, J.: *History of Indian and Eastern Architecture*, Radhey Shayam Press, Poona. 1967.

Frauwallner, E.: *History of Indian Philosophy*, Motilal, Delhi, 1973.

Garbe, R.: *The Philosophy of Ancient India*, Chicago University Press, Chicago, 1899.

Gopinath, K.: *Aspects of Vedic Philosophy*, University of Burdwan, Calcutta, 1966.

Grisenold, H.D.: *Insights into Modern Hinduism,* Adavita Ashrama, Calcutta, 1977.

Hackett, S.: *Oriental Philosophy: A Westerner's Guide to Eastern Thought,* University of Wisconsin Press, Madison, 1979.

Isaeva, V.: *Shankara and Indian Philosophy*, SUNY, New York, 1993.

Jain, Girilal: *The Hindu Phenomenon*, UBSPD, Delhi 1994.

John, G.: *The New Technology and Human Values*, Bangiya Sahitya Parishad Patrika (Bengali), Calcutta, 1901.

Karl, H. : *Advaita Vedanta up to Sankara and his Pupils: Encyclopedia of Indian Philosophies*, Princeton University Press, Princeton, 1981.

Knappert, J.: *Indian Mythology*, Saraswati Press, Delhi, 1974.

Kuiper, F.B.J.: *Aryans in the Rigveda*, Rodopi, Amsterdam, 1991.

Lal, M.: *The Ramayana of Valmiki*, Munshiram Manoharlal, New Delhi, 1978.

Mahadevan, T.M.P.: *Philosophy of Advaita*, Luzac & Company, London, 1938.

Mahalingam, T.V.: *Administration and Social Life under Vijayanagar*, University of Madras, Madras, 1969.

Mathew, A.: *Culture and Anarchy*, Kali Ram Press, Calcutta, 1962.

Mokashi, D.B.: *Palkhi: An Indian Pilgrimage*, State University of New York Press, Albany, 1990.

Narain, H.: *Myths of Composite Culture and Equality of Religions*, Voice of India, Delhi 1991.

Narasimhan, C.V.: *The Mahabharata: An English Version Based on Selected Verses*, Columbia University Press, New York, 1965.

Oswald, S.: *Man and Technics: A Contribution to Philosophy of Life*, MacMillan, New York, 1932.

Pagal, B.: *Temple of the Phallic King: The Mind of India*, Simon and Schuster, New York, 1998.

Potter, H.: *Encyclopedia of Indian Philosophies*, Princeton University Press, Princeton, 1981.

Pushpendra, K.: *Hindu Religion and Ethics,* Asian Publication Services, Delhi, 2002.

Ramamurti, R.: *Advaitic Mysticism of Sankara*, Theosophical Publishing House, Wheaton 1967.

Ranganathananda, S.: *Science and Religion,* Routledge and Kegan Paul, London, 1970.

Reuchenbach, H.: *The Philosophy of Space and Time*, Dover Publications, New York, 1958.

Rose, J.: *The Cybernetic Revolution*, Advaita Ashram, Calcutta, 1978.

Sharma, L. P.: *The Vedic Philosophy*, MacMillan, New Delhi, 1988.

Shashibhusan, D.: *An Introduction to Vedic Philosophy*, Sterling Publishers, New Delhi, 1981.

Subramaniam, Kamala: *Mahabharata*, Rawat Press, Bombay, 1971.

Tarkalankara, C.: *Hindu Philosophy,* Basu Mallik Press, Calcutta, 1922.

Taylor, E.: *Elements Metaphysics*, Oxford University Press, London, 1952.

Tejomayananda, S.: *Hindu Culture: An Introduction,* Sage, Walnut Creek, 1995.

Thomas, R.: *Kautilya and the Arthasastra: A Statistical Study*, Brill, Leiden, 1971.

Utpat, V.N.: *Riddles of Buddha and Ambedkar,* Itihas Patrika Prakashan, Thane 1988.

Vasant, G.: *Mysterious Kundalini*, George Allen and Unwin, London, 1961.

Visvantha, V.: *Essentials of Vedic Philosophy,* Narosa Pub. House, Delhi, 1998.

Vivekananda, S.: *Raja Yoga,* Chinmaya Publications, Piercy, 1993.

Warrier, K.: *Srimad Bhagavad Gita Bhashya of Sri Sankaracarya*, B. I. Publications, Bombay, 1973.

Wendy, D.: *Laws of Manu*, Penguin, New York, 1992.

Wiener, N.: *God and Golem*, Methnen & Co., London, 1924.

William, G.: *Indian and the Future*, MacMillan, London. 1917.

Wilson, T.: *The Self in Indian Philosophy*, Mounton & Co., London, 1964.

Yong, C.: *Oriental Thought: An Introduction to the Philosophical and Religious Thought of Asia,* Rowman and Allanheld, Totowa, 1981.

Zimmer, H.: *Philosophies of India*, Joseph Campbell. New York, 1951.

# Index

## C



## D



## E



## F



## G

## V



## W



## Y



## Z



❑❑❑